ROUTLEY-MEYER TERNARY RELATIONAL SEMANTICS FOR INTUITIONISTIC-TYPE NEGATIONS

ROUTLEY-MEYER TERNARY RELATIONAL SEMANTICS FOR INTUITIONISTIC-TYPE NEGATIONS

GEMMA ROBLES

JOSÉ M. MÉNDEZ

ACADEMIC PRESS
An imprint of Elsevier

Library of Congress Cataloging-in-Publication Data
A catalog record for this book is available from the Library of Congress

British Library Cataloguing-in-Publication Data
A catalogue record for this book is available from the British Library

ISBN: 978-0-08-100751-8

For information on all Academic Press publications
visit our website at https://www.elsevier.com/books-and-journals

Working together
to grow libraries in
developing countries

www.elsevier.com • www.bookaid.org

Publisher: Mathew Deans
Acquisition Editor: Glyn Jones
Editorial Project Manager: Serena Castelnovo
Production Project Manager: Surya Narayanan Jayachandran
Designer: Mark Rogers

Typeset by VTeX

CONTENTS

ABOUT THE AUTHORS

Gemma Robles is a researcher at the Department of Psychology, Sociology, and Philosophy at the Universidad de León. Since 2001, she has published more than 50 papers on non-classical logics in impact journals.

José M. Méndez is a Professor of Logic at the Universidad de Salamanca. His research interests include philosophical logic focusing on modal logics, multivalued logics, and relevance logics.

PREFACE

Routley-Meyer type ternary relational semantics (henceforward, RM-semantics) is a general semantics for non-classical logics. There are essentially two types of RM-semantics that can be dubbed here RM_1-semantics and RM_0-semantics (cf. §0.1 below). The basic positive logics (i.e., without negation) B_+ and B_{K+} are the minimal positive logics interpretable with RM_1-semantics and RM_0-semantics, respectively.

In RM-semantics in general, negation is interpreted by means of the Routley operator, which is adequate for modelling only De Morgan-type negations and extensions thereof.

The fundamental reference works on RM-semantics are still [37] and [9] and, specially, Chapter 4 of [37], where it is shown how to define a wealth of extensions of Routley and Meyer's basic logic B. The logic B is a De Morgan negation expansion of the basic positive logic B_+, also defined by the referred authors.

The purpose of the present work is to mirror Chapter 4 of [37] from the perspective of intuitionistic-type negations, instead of De Morgan-type negations. As it is known, negation can be rendered as follows according to the intuitionistic view: No-X is assertable if and only if X entails the Absurd (cf. the heads "Intuitionistic Logic" and related topics in [16] and [44]).

In particular, we aim at (1) to define the basic logics expanding B_+ and B_{K+} with intuitionistic-type negations; and (2) to define in RM_1-semantics and RM_0-semantics, as the case may be, a wealth of extensions of the basic systems, similarly as in Chapter 4 of [37], but with intuitionistic-type negations this time.

The structure of the present work is as follows. *Introduction*: We explain the general characteristics of RM-semantics. Also, how intuitionistic-type negations can be introduced in it by following the pattern according to which this type of negations is defined in standard positive binary Kripke semantics. *Part 1*: The basic positive logics B_+ and B_{K+} are recalled and the basic logics expanding them with intuitionistic-type negations are introduced. *Part 2*: There are four subparts. *Chapter 4*: We alternatively define by means of a falsity constant the basic logics introduced in Part 1 where they were defined by using the unary negation connective. *Chapter 5*: Two different versions of what can be named the basic constructive relevance logic are provided since the logics studied in Part 1 are not relevance logics. *Chapter 6*: We generally define extensions and expansions of the basic logics introduced in Chapters 1–5. Finally, in Chapter 7, the reader can find a brief discussion on some of the extensions and expansions defined in the preceding chapters.

To the best of our knowledge, a project like the present one cannot be found in the literature. There are some particular studies (cf., e.g., [22] and [38]), most of them by us, by the way (cf., as a way of an example, [20], [25], and [31]), some general studies on constructive negations or on the relations between intuitionism and relevance (cf., e.g., [21], [41]) or the well-known treatise on relational semantics for non-classical logics [7], but not a general development of intuitionistic-type negations in RM_1 and RM_0 semantics. Furthermore, it has to be remarked that this type of negations is not considered in the two volumes of *Entailment* (cf. [2], [3]), while in the case of the two volumes of *Relevant Logics and Their Rivals* (cf. [9] and [37]), intuitionistic-type negations are summarily treated in the four pages constituting Chapter 6 of [9]. In some sense, the pages that follow can be considered as a development of the referred chapter, although it has to be noted that the second author of the present work was writing on intuitionistic-type negations in RM-semantics in 1987 (cf., e.g., [19]).

We have tried to write a self-contained book and in fact we think that it can be read by anyone having the fundamental notions on classical and modal propositional logics.

We thank the Spanish government which has funded the research projects by the second author and Francisco Salto since 1991 and those by both authors and Francisco Salto since 2001. Currently, we are supported by the research project FFI2014-53919-P. Some of these projects were dedicated to the study of intuitionistic-type negations in RM-semantics (cf. http://campus.usal.es/~glf/). In addition, the first author also thanks the Spanish government for the funding of her Juan de la Cierva and Ramón y Cajal projects, the latter being dedicated to intuitionistic-type negations (cf. http://grobv.unileon.es/).

Finally, we dedicate this book to our respective parents: Manolita Vázquez Terrón and Blas Robles Díaz, and Lola Rodríguez de la Peña and Marcos Méndez Sastre (*in memoriam* (1923–2010)).

INTRODUCTION

0.1 TERNARY RELATIONAL SEMANTICS. GENERAL CHARACTERISTICS

Routley-Meyer type ternary relational semantics (RM-semantics) was introduced by Richard Routley and Robert K. Meyer in the early seventies of the past century (cf. [33–35]). RM-semantics was particularly defined for interpreting relevance logics, but it was soon noticed that an ample class of logics not belonging to the relevance logics family could also be characterized by this semantics. Actually, it was verified that RM-semantics is a highly malleable semantics capable of modelling families of logics which are very different from each other (cf. [37], [9], and the references quoted in both volumes). As it has been noted above, the fundamental reference works in RM-semantics are [37] and [9], and specially Chapter 4 of [37]. On the other hand, it has to be remarked that A. Urquhart [42] and K. Fine [15] presented similar semantics to RM-semantics around the same time the latter was introduced and with the same aim: to endow relevance logics with a semantics (some brief historic notes on this question can be read in [3] §48; cf. also the general project on the topic in [6] and the special issue [8]).

RM-semantics is a relational type semantics. It can be distinguished from standard Kripke semantics in two fundamental aspects. (a) As patently indicated in its name, the accessibility relation between worlds (points, "set-ups" or whatever the name is preferred) is a ternary relation instead of a binary one, as it is the case in standard Kripke semantics. (b) Negative formulas are interpreted by the Routley unary operator (or "Routley star") (cf. [36], [37]) in each possible world w.r.t. its so called "star-image world" ("*-image world") instead of being interpreted in each possible world in function of the argument's value of the negation formula in that same possible world, as it is the case in standard Kripke semantics.

There are essentially two types of RM-semantics that can be dubbed here RM_1-semantics and RM_0-semantics. RM_1-semantics is RM-semantics with a set of designated points w.r.t. which validity of formulas is decided; RM_0-semantics is RM-semantics without this set. In RM_0-semantics validity of formulas is decided w.r.t. the set of all points. These two types of semantics can also be found in standard binary Kripke semantics.

In the following pages, we shall comment on these characteristics of RM-semantics in order to establish a context within which our RM-semantics for intuitionistic-type negations can be developed.

0.2 POSITIVE MODELS. THE INTERPRETATION OF THE CONDITIONAL

We shall understand here the term "relevance logic" in its minimal sense: a logic L is a "relevance logic" if it has the "variable-sharing property" (vsp), that is, if in all its theorems of conditional form antecedent and consequent share at least a propositional variable. Consequently, a "paradox of relevance" should be understood as a conditional in which antecedent and consequent do not have variables in common (cf. [2]).

The first thing one realizes when trying to define a relational semantics for relevance logics (there are other options like, say, algebraic semantics) is that standard binary relational semantics is inadequate. Consider a positive (i.e. without negation) language (cf. Definition 1.1 below) with the connectives → (conditional), ∧ (conjunction), and ∨ (disjunction). Standard binary relational semantics with a set of designated points is defined for this language as follows.

Definition 0.1 (Positive Kripke models with a set O). A positive binary model with a set of designated worlds, O, is a structure $(K, O, R, \vDash)$ where (i) O is a non-empty subset of K (K is the set of all possible worlds; O is the set of all designated possible worlds —possible worlds with some special properties); (ii) $R \subseteq K^2$ and R has certain properties (R is a binary relation on K with some determined properties: it is the accessibility relation between possible worlds); and (iii) $\vDash \subseteq K \times \mathcal{F}$ and it is such that for each $p \in \mathcal{P}$, $A, B \in \mathcal{F}$, and $a \in K$, the following conditions (clauses) are fulfilled: ($\vDash$ is a (valuation) relation verifying or falsifying each wff in each possible world according to the following conditions)

(1). $a \vDash p$ or $a \nvDash p$
(2). $a \vDash A \wedge B$ iff $a \vDash A$ and $a \vDash B$
(3). $a \vDash A \vee B$ iff $a \vDash A$ or $a \vDash B$
(4). $a \vDash A \to B$ iff for all $b \in K$, $(Rab \ \& \ b \vDash A) \Rightarrow b \vDash B$

On the other hand, the definition of validity is as follows $\vDash A$ iff $a \vDash A$ for all $a \in O$ in all positive binary models with a set of designated worlds (A is valid iff A is true in each designated world in all positive binary models with a set of designated worlds) (cf. Definition 1.4 below).

We consider that the definition of a class of models together with the definition of validity, given that class of models, determines a semantics (cf. Definition 1.7 below).

Notice that, given a positive binary model with a set O and, according to condition (4), a formula of conditional form is true in a possible world a iff in all accessible worlds from a in which the antecedent is true, the consequent is also true.

Next, it is shown that the standard binary relational semantics just defined validates (no matter the clause for negation added to it) paradoxes of relevance such as the characteristic S4 axiom $B \rightarrow (A \rightarrow A)$ (cf. [18]).

Proposition 0.2 (Validity of $B \rightarrow (A \rightarrow A)$). *The thesis $B \rightarrow (A \rightarrow A)$ is valid in the binary relational semantics defined in Definition 0.1.*

Proof. For reductio, suppose that there are $A, B \in \mathcal{F}$ and $a \in O$ in some model such that (1) $a \nvDash B \rightarrow (A \rightarrow A)$. Then, there is $b \in K$ in this model such that (2) Rab, $b \vDash B$, and $b \nvDash A \rightarrow A$. Given $b \nvDash A \rightarrow A$, there is $c \in K$ in this model such that (3) Rbc, $c \vDash A$, and $c \nvDash A$. But the situation described in (3) is impossible. $\quad\square$

Consequently, the thesis $B \rightarrow (A \rightarrow A)$ is valid in the binary relational positive semantics with a set of designated points outlined in Definition 0.1, whence it must be concluded that no relevance logic can be characterized by this semantics (notice that $B \rightarrow (A \rightarrow A)$ is valid independently of the properties predicable of the accessibility relation).

As we are going to see immediately, a first step for generally invalidating the paradoxes of relevance is to change the binary relation for a ternary one. The basic idea is to change condition (4) (in Definition 0.1) for condition (4′) which reads

$$(4') \; a \vDash A \rightarrow B \text{ iff for all } b, c \in K, (Rabc \; \& \; b \vDash A) \Rightarrow c \vDash B$$

In this way, positive RM_1-semantics is defined as follows (cf. §0.1):

Definition 0.3 (Basic positive ternary models with a set O). A basic positive ternary model with a set of designated worlds, O, is a structure $(K, O, R, \vDash)$ where $R \subseteq K^3$ and K, O, and $\vDash$ are defined similarly as in Definition 0.1 except that clause (4) is changed for clause (4′).

Notice that a formula of conditional form is now true in a possible world a iff in all worlds b, c such that $Rabc$ the consequent is true in c if the antecedent is true in b. (On the interpretation of the relation R, cf. [5], [6], [8], and [37].)

Concerning the paradoxes of relevance, the point is that now $A \to A$ can be false in some possible worlds. Let us see it. Suppose $A \to A$ is not true in a. This only means that there are $b, c \in K$ such that $Rabc$, $b \vDash A$, and $c \nvDash A$. But $b \vDash A$ and $c \nvDash A$ are not, of course, contradictory situations if b and c are different worlds (nevertheless, $A \to A$ is true in *all designated* worlds). Consider now $a \in O$ and $b, c \in K$ such that $Rabc, b \vDash B$, and $c \nvDash A \to A$. In principle, there is nothing absurd in this supposition. Then, $B \to (A \to A)$ is not true in a, whence this formula is not valid in the semantics in Definition 0.3.

As we shall see, a wealth of structures built upon the semantics sketched in Definition 0.3 characterize relevance logics. However, it has to be remarked that the ternary relation is not *per se* sufficient for invalidating the paradoxes of relevance. Suppose that the set O is dropped and consequently validity is defined w.r.t. K: $\vDash A$ iff $a \vDash A$ for all $a \in K$ in all models (A is valid iff A is true in all points in all models). In this case, we have positive RM_0-semantics that are defined as follows (cf. §0.1):

Definition 0.4 (Basic positive ternary models without the set O). A basic positive ternary model without O is a structure $(K, R, \vDash)$ where K, R, and $\vDash$ are defined exactly as in Definition 0.3 (K is non-empty), and validity is defined w.r.t. K as indicated above.

We show that paradoxes of relevance are included in the set of valid formulas of these models by proving that the rule Veq preserves validity.

Proposition 0.5 (The rule Veq preserves validity). *The rule Veq ("Verum e quodlibet" —"A true proposition follows from any proposition"), $A \Rightarrow B \to A$, preserves validity in the positive ternary semantics outlined in Definition 0.4.*

Proof. For reductio, suppose that there are $A, B \in \mathcal{F}$ such that (1) $\vDash A$ but (2) $\nvDash B \to A$. Then, there is $a \in K$ in some model such that (3) $a \nvDash B \to A$. Given 3, there are $b, c \in K$ in this same model such that (4) $Rabc$, (5) $b \vDash B$, and (6) $c \nvDash A$. But 1 and 6 contradict each other. Consequently, Veq preserves validity in the positive ternary semantics summarily described in Definition 0.4: if A is valid, then $B \to A$ is valid, for any $B \in \mathcal{F}$.   □

Let now A be a valid wff and B a wff with no propositional variables in common with A. By Proposition 0.5, $B \to A$ is valid, but $B \to A$ is a paradox of relevance (A and B share no propositional variables). Consequently, no ternary relational semantics with validity defined w.r.t. K can characterize a relevance logic.

We note that B_+-models determining the minimal positive logic B_+ definable in RM_1-semantics (RM-semantics with a set of designated points) are easily defined from Definition 0.3 (cf. Definition 1.12). On the other hand, B_{K+}-models determining the logic B_{K+}, the minimal logic in RM_0-semantics (RM-semantics without a set of designated points), are easily definable from Definition 0.4 (cf. Definition 3.6).

In sum, it follows from the discussion developed above that a relational semantics needs both a ternary relation and a set of designated worlds w.r.t which to define validity, in order to expel paradoxes of relevance from the set of valid formulas.

However, the present work is not focused on relevance logics, but on logics endowed with intuitionistic-type negations that are representable in RM-semantics. From this point of view, the advantage of positive RM-semantics over standard binary semantics is simply the much wider range of logics the former semantics can interpret. As remarked above, B_+ and B_{K+} are the minimal positive logics definable in RM_1-semantics and RM_0-semantics, respectively. The logics B_+ and B_{K+} are weak logics that can be variously extended in several divergent directions, most of which are modelable in RM_1-semantics or RM_0-semantics, as the case may be. None of the extensions of B_{K+} is a relevance logic, but many of B_+ are (cf. Chapters 6 and 7). However, consider now, on the other hand, the binary semantics outlined in Definition 0.1. The question is which is the minimal logic this semantics characterizes. Suppose that possible worlds are canonically interpreted as theories (sets of formulas with certain properties). And suppose that the accessibility relation is canonically interpreted via the conditional as follows: for any theories a, b, Rab iff $(A \rightarrow B \in a \ \& \ A \in b) \Rightarrow B \in b$. That is, b is accessible from a iff the consequent of each conditional appearing in a is present in b provided the antecedent is also present in b. Then, although this cannot be proved here, the answer is that this minimal logic is the positive fragment of Lewis' S4, $S4_+$, as axiomatized by Hacking in [18] (cf. [30]). But $S4_+$ is a strong logic that only can be extended through foreseeable paths leading, for example, to $S5_+$, or to propositional intuitionistic logic H_+.

0.3 THE INTERPRETATION OF NEGATION IN RM-SEMANTICS

As we have seen, a ternary accessibility relation together with a set of designated worlds w.r.t. which validity is defined seems, in principle, sufficient to maintain paradoxes of relevance outside the set of valid formulas in RM-semantics. Concerning negation, the first point to remark is that the standard (boolean) condition of binary semantics is inadequate. For, suppose that condition (v) that follows is added to the models in Definition 0.3

(v) For all $A \in \mathcal{F}$ and $a \in K$, $a \vDash \neg A$ iff $a \nvDash A$

It is clear that the axiom Ecq ("E contradictione quodlibet" —"Any proposition follows from a contradiction") is immediately validated.

Proposition 0.6 (Validation of the axiom Ecq). *The axiom Ecq $(A \wedge \neg A) \to B$ is valid in the semantics sketched in Definition 0.3 supplemented with the condition (clause) (v) recorded above.*

Proof. Suppose that there are $A, B \in \mathcal{F}$ and $a \in O$ in some model such that (1) $a \nvDash (A \wedge \neg A) \to B$. Then, there are $b, c \in K$ in this model such that (2) $Rabc$, $b \vDash A \wedge \neg A$, and $c \nvDash B$. Given (2), we have (3) $b \vDash \neg A$ and $b \vDash A$. But the situation in (3) is impossible according to clause (v). $\square$

Consequently, introducing negation via clause (v) in the models in Definition 0.3 amounts to introduce paradoxes of relevance, so laboriously eliminated precisely by the definition of such models.

The idea for adequately introducing negation in the models in Definition 0.3 is owed to R. Routley and V. Routley (née Plumwood) (cf. [36], [37] and references therein). These authors propose to interpret negative formulas by means of a unary operator (currently labelled "Routley operator" or "Routley star") according with the following condition:

(v'). $a \vDash \neg A$ iff $a^* \nvDash A$

That is to say, a negative formula is true in a possible world iff its argument is false in the corresponding "star-image world" or "*-image world".

This interpretation of negation has been much discussed (cf. [12], [23]), although it is to be noted that it has precedents in the algebraic approach to logic of the Polish school (cf. [3], §18.2). Be it as it may (we cannot pause to discuss this question here), it is easy to see that the axiom Ecq is no longer valid if clause (v) is changed for clause (v'): given line (3) of Proposition 0.6, we have now (4) $b^* \nvDash A$ and $b \vDash A$, which is not a contradictory statement at all. Actually, (3) and (4) reveal that possible worlds can be inconsistent in RM-semantics when negation is interpreted according to clause (v'). But they can also be incomplete: given $a \nvDash \neg A$ and $a \nvDash A$ for some $a \in K$ in some model, $a^* \vDash A$ and $a \nvDash A$ follow, a non-contradictory situation again (hence the preference, on the part of the creators of RM-semantics, for the term "set-up" over "possible-world").

Two remarks to end the section. (a) Clause (v) has been added to ternary models based upon the basic ones in Definition 0.3 with unexpected results. Maybe the more conspicuous of these is that addition of clause (v) to models characterizing relevance logic R (cf. [33], [37]) does not cause the collapse of R in classical propositional logic (cf. [37]). (b) The second remark has to do with the minimal logic characterized by RM-semantics when negation is interpreted following clause (v'). Roughly speaking, the semantics characterizing B_+ (outlined in Definition 0.3) supplemented with clause (v') (and some basic postulates governing *) determines Sylvan and Plumwood's logic B_M (cf. [40]), which is the result of adding a minimal De Morgan negation to B_+. In particular, this minimal De Morgan negation is axiomatized by adding the following axioms and rule: $(\neg A \wedge \neg B) \to \neg (A \vee B)$, $\neg (A \wedge B) \to (\neg A \vee \neg B)$ and Contraposition, $A \to B \Rightarrow \neg B \to \neg A$. But minimal as it is, this De Morgan negation is not, of course, an intuitionistic-type negation, whence it has to be concluded that the latter type of negation has to be introduced in RM-semantics by an expedient different from the Routley operator. (Nevertheless, we note that this minimal De Morgan negation —and stronger ones— can be interpreted in the models we propose —cf. Chapter 6.)

0.4 THE INTRODUCTION OF INTUITIONISTIC-TYPE NEGATION IN STANDARD POSITIVE BINARY SEMANTICS

As pointed out above, from an intuitionistic point of view, a negation $\neg A$ is assertable iff A entails the Absurde (cf. the heads "Intuitionistic logic" and related topics in [16] and [44]). Let us formally represent the Absurde by a propositional falsity constant f. Then, intuitionistic-type negation can be introduced by a falsity constant f (along with the definition $\neg A = A \to f$) or by means of the unary connective $\neg$. Suppose negation is introduced by f. We have to consider two possibilities: (a) f can hold in some possible worlds; (b) f fails in every possible world. Let us use S to refer to the subset of the set of all possible worlds in which f does not hold (S is, so to speak, the set of all "consistent worlds"). Consider now an expansion with f of the positive language discussed above. Kripke models for this expanded language can be defined as follows (cf. [13], [14]).

Definition 0.7 (Kripke models with O for f). A binary model with a set of designated worlds, O, for the positive language plus the falsity constant f is a structure $(K, O, S, R, \vDash)$ where $S \subseteq K$ and $O \cap S \neq \emptyset$, and K, O, R, and $\vDash$ are defined exactly as in Definition 0.1, except for the addition of one of the clauses (5) or (5′) that follow:

$$(5) \text{ For all } a \in K, a \vDash f \text{ iff } a \notin S$$
$$(5') \text{ For all } a \in K, a \nvDash f$$

The clause (5) is added if, according to (a) above, f can be true in some possible worlds. The clause (5′) is added if, as stated in (b), f holds in no possible world (in this case, $S = K$ and the set S can be deleted in the models. Notice that the condition $O \cap S \neq \emptyset$ states that there are designated worlds that are consistent).

Now, more than in the models just defined, we are interested in the following question: which is the negation clause based upon (5) and which is the negation clause based upon (5′), when negation is defined via f (i.e., $\neg A = A \to f$), as pointed out above?

(i) *The negation clause according to (5)*: For any model $a \in K$, we have $a \vDash \neg A$ iff $a \vDash A \to f$ (definition of $\neg$). Then, by applying clause (4) (cf. Definition 0.1), we get $a \vDash \neg A$ iff for all $b \in K$, $(Rab \ \& \ b \vDash A) \Rightarrow b \vDash f$. Finally, by clause (5), we obtain $a \vDash \neg A$ iff for all $b \in K$, $(Rab \ \& \ b \in S) \Rightarrow b \nvDash A$. Consequently, the negation clause according to (5) is:

$$(5\neg). \ a \vDash \neg A \text{ iff for all } b \in K, (Rab \ \& \ b \in S) \Rightarrow b \nvDash A$$

(ii) *The negation clause according to (5′)*: By using clause (5′) similarly as (5) was used in (i), we have that the negation clause according to (5′) is the following:

$$(5'\neg). \ a \vDash \neg A \text{ iff for all } b \in K, Rab \Rightarrow b \nvDash A$$

In this way, binary models for the expansion of the positive language with the unary connective $\neg$ can be defined upon the models in Definition 0.1 as follows.

Definition 0.8 (Kripke models with O for $\neg$). A binary model with a set of designated worlds, O, for the positive language plus the unary connective $\neg$ is a structure $(K, O, S, R, \vDash)$ where $S \subseteq K$ and $O \cap S \neq \emptyset$, and $K, O, R,$ and $\vDash$ are defined exactly as in Definition 0.1, except for the addition of one of the clauses (5$\neg$) or (5$'\neg$) together with (possibly) some semantical postulates expressing some additional properties of R. (If (5$'\neg$) is added, then $S = K$ and the set S can be deleted in the models.)

In intuitionistic logic the Absurde is canonically interpreted as any contradictory statement, that is, any formula of the form $A \wedge \neg A$. This interpretation is adequate for some of the logics defined in Chapters 6 and 7, but not for the basic logics introduced in the first five chapters and many of their extensions defined in the last two chapters. Therefore, in the present work, the Absurde is generally interpreted as the negation of a theorem (of the logic in question), from a canonical point of view. Thus, inconsistency is here understood as the presence of the negation of a theorem, consistency being the absence of inconsistency in the explained sense. This concept of consistency has been labelled "weak-consistency" (w-consistency) (cf. [28] and references therein), and, as suggested, is central in the development of the pages that follow. Suppose now that this interpretation of the Absurde and the associate notion of w-consistency is adopted, then, albeit we cannot pause here to prove it, we remark that the semantics outlined in Definition 0.8 with clause (5$\neg$) added characterizes the system resulting from adding the following axiom and rule (cf. Chapter 2 below) to S4$_+$: A7 $\neg A \rightarrow [A \rightarrow \neg(B \rightarrow B)]$; Red $A \Rightarrow (B \rightarrow \neg A) \rightarrow \neg B$. On the other hand, if (5$'\neg$) is the clause added, the system characterized is the result of adding to S4$_+$ the same rule but the axiom A8 $\neg A \rightarrow (A \rightarrow B)$ instead of A7 (cf. [30]).

0.5 THE INTRODUCTION OF INTUITIONISTIC-TYPE NEGATION IN RM-SEMANTICS

The introduction of intuitionistic-type negation in RM-semantics mirrors the introduction of the same type of negation in standard binary semantics as it has been reviewed in the precedent section. Thus, we will begin with the following definition.

Definition 0.9 (Ternary models with O for f). A ternary model with a set of designated worlds, O, for the positive language plus the falsity constant f is a structure $(K, O, S, R, \vDash)$ where $S \subseteq K$ and $O \cap S \neq \emptyset$, and $K, O, R,$ and

⊨ are defined exactly as in Definition 0.3 except for the addition of one of the clauses (5) or (5') formulated in Definition 0.8. As it was the case there, clause (5) is added if f can be true in some worlds, while clause (5') is the one added when f fails in each possible world (in the latter case $S = K$ and consequently the set S can be deleted in the models).

Given the models in Definition 0.9, we proceed to establish the negation clauses corresponding to clauses (5) and (5') in RM-semantics.

(i) *The negation clause according to (5)*: For any model and $a \in K$, we have $a \vDash \neg A$ iff $a \vDash A \rightarrow f$ (Definition of $\neg$). Then, by applying clause (4') (cf. Definition 0.3), we get $a \vDash \neg A$ iff for all $b, c \in K$, $(Rabc \ \& \ b \vDash A) \Rightarrow c \vDash f$. Finally, by clause (5), we obtain $a \vDash \neg A$ iff for all $b, c \in K$, $(Rabc \ \& \ c \in S) \Rightarrow b \nvDash A$. Consequently, the negation clause according to (5) is:

$$(5\neg) \ a \vDash \neg A \text{ iff for all } b, c \in K, \ (Rabc \ \& \ c \in S) \Rightarrow b \nvDash A$$

(ii) *The negation clause according to (5')*: A similar proof to the precedent one based now upon clause (5') shows that the negation clause according to (5') is the following:

$$(5'\neg) \ a \vDash \neg A \text{ iff for all } b, c \in K, \ Rabc \Rightarrow b \nvDash A$$

On the other hand, ternary models built upon Definition 0.4 can be defined as follows:

Definition 0.10 (Ternary models without O for f). A ternary model without O for the positive language plus f is a structure $(K, S, R, \vDash)$ where $S \subseteq K$ and $S \neq \emptyset$, and K, R, and $\vDash$ are defined exactly as in Definition 0.4 except for the addition of one of the clauses (5) or (5') formulated in Definition 0.7. (If (5') is added, then $S = K$ and consequently the set S can be deleted in the models.)

Then, ternary models for the expansion of the positive language with the unary connective $\neg$ can be defined upon the models in Definition 0.3 as follows:

Definition 0.11 (Ternary models with O for $\neg$). A ternary model with a set of designated worlds, O, for the positive language plus the unary connective $\neg$ is a structure $(K, O, S, R, \vDash)$ where $S \subseteq K$ and $O \cap S \neq \emptyset$, and K, O, R, and $\vDash$ are defined exactly as in Definition 0.3 except for the addition of one of the clauses $(5\neg)$ or $(5'\neg)$ together with (possibly) some semantical postulates expressing some additional properties of R. (If $(5'\neg)$ is added, then $S = K$ and consequently the set S can be deleted in the models.)

On the other hand, ternary models built upon Definition 0.4 can be defined as follows:

Definition 0.12 (Ternary models without O for $\neg$). A ternary model without O is a structure $(K, S, R, \vDash)$ where $S \subseteq K$ and $S \neq \emptyset$, and K, R, and $\vDash$ are defined exactly as in Definition 0.4 except for the addition of one of the clauses $(5\neg)$ or $(5'\neg)$ together with (possibly) some semantical postulates expressing some additional properties of R. (If $(5'\neg)$ is added, then $S = K$ and consequently the set S can be deleted in the models.)

The present work is dedicated to investigate the logics characterized by the semantics in Definitions 0.9–0.12 together with their extensions characterized in their turn by restrictions in the semantics outlined in these definitions. We remark that the additional elements required for the models in Definitions 0.9–0.12 to define the minimal logics in the appropriate language are provided in Definitions 4.1, 2.2, and 3.24, respectively. These definitions characterize the logics $B_{+,f}$, $B_{K+,f}$, B_{cS}, and B_{KS}, respectively. If the set S is deleted, the referred additional elements can be found in Definitions 2.30 and 3.32 characterizing the logics B_c and B_K, respectively (cf. also Proposition 6.16). The logics $B_{+,f}$, B_{cS}, and B_c (respectively, $B_{K+,f}$, B_{KS}, and B_K) are expansions of the positive logic B_+ (respectively, B_{K+}), referred to above. These basic logics just mentioned can be axiomatized similarly as the logics expanding $S4_+$ characterized by Definition 0.8 were axiomatized (cf. §2.6).

A final note to end this introduction. The basic logics in Definitions 0.11 and 0.12 are not relevance logics, since they contain paradoxes of relevance. Nevertheless, it is possible to define significant relevance logics with intuitionistic-type negations introduced according to clause $(5\neg)$ (cf. Chapter 5).

PART 1

CHAPTER 1

The basic positive logic B_+. EB_+-models

In this chapter, we define the basic positive logic B_+ and the notion of an EB_+-model for interpreting extensions and expansions of B_+. Also, we establish the basic facts for proving the soundness and completeness of B_+ and of extensions and expansions of it. In the last section, we discuss some notions of importance in the book: "rule of inference", "rule of proof" and the appropriate concept of "strong completeness" for the logics presented in the pages to follow.

1.1 EB_+-MODELS

We need some preliminary notions.

Definition 1.1 (Languages). We shall consider propositional languages with a denumerable set of propositional variables $p_0, p_1, ..., p_n, ...$ and at least the connectives $\rightarrow$ (conditional), $\wedge$ (conjunction), and $\vee$ (disjunction). Additional connectives are $\neg$ (negation) and the propositional constants t (truth) and f (falsity) (propositional constants can be viewed as 0-ary connectives). The biconditional ($\leftrightarrow$) and the set of wffs are defined in the customary way. A, B etc. are metalinguistic variables. By $\mathcal{P}$ and $\mathcal{F}$ we shall refer to the set of all propositional variables and the set of all wffs of each one of the languages considered, respectively.

Definition 1.2 (Logics). A logic L is a structure $(\mathcal{L}, \vdash_L)$ where $\mathcal{L}$ is a propositional language and $\vdash_L$ is a (proof-theoretical) consequence relation defined in $\mathcal{L}$ by a set of axioms and a set of rules of derivation. The notions of 'proof' and 'theorem' are understood as it is customary in Hilbert-style axiomatic systems ($\vdash_L A$ means that A is a theorem of L).

Definition 1.3 (Extensions and expansions of a logic L). Let L be a logic formulated with axioms $a1, ..., an$ and rules of derivation $r1, ..., rm$. A logic L' includes L iff $a1, ..., an$ are theorems of L' and $r1, ..., rm$ are provable in L'. We shall generally refer to logics including L by EL-logics. Notice that an EL-logic can be an extension of L (a strengthening of L in the

Routley–Meyer Ternary Relational Semantics for Intuitionistic-Type Negations.
DOI: http://dx.doi.org/10.1016/B978-0-08-100751-8.00002-X

language of L) or an expansion of it (a strengthening of L in an expansion of the language of L). An extension L′ of L is a proper extension if L′ is not included in L.

In the present work, basic logics with intuitionistic-type negations are built upon two different minimal positive logics, Routley and Meyer's basic positive logic B_+ (cf. [35] and [37]) on the one hand, and the logic B_{K+}, defined by the authors of this book (cf. [32], [27] —cf. also [26]), on the other. B_+ is a relevance logic, but B_{K+} (axiomatized when adding to B_+ the rule Veq or rule K —$A \Rightarrow B \to A$) does not have the "variable-sharing property" (the logics B_+ and B_{K+} are defined below in Definitions 1.10 and 3.1, respectively). B_+ is characterized by the simplest RM_1-semantics (RM-semantics with a set of designated points O), whereas B_{K+} is determined by the simplest RM_0-semantics (RM-semantics without this set O). Thus, although B_{K+} is an extension of B_+, it will be useful to distinguish between EB_+-models, which are semantic structures introduced for interpreting extensions and expansions of B_+ requiring the presence of a set of designated points O, and EB_{K+}-models, designed for interpreting extensions and expansions of B_{K+}, which do not need the set O. Notice, however, that by an EB_+-logic we generally refer to an extension or an expansion of B_+, including the extensions and expansions of B_{K+}.

Next, the notion of an EB_+-model is defined; that of an EB_{K+}-model is introduced in Definition 3.3. Except for the truth constant t that will be used for defining in Chapter 5 relevance logics with intuitionistic-type negations, all elements in EB_+-models (and in EB_{K+}-models) have been discussed in the introduction.

Definition 1.4 (EB_+-models). An EB_+-model M is a structure with at least the following items (1) A set K and a subset of it, O; (2) a ternary relation R defined on K subject to the following definitions and postulates for all $a, b, c, d \in K$:

d1. $a \leq b =_{df} \exists x \in O\ Rxab$

d2. $R^2abcd =_{df} \exists x \in K(Rabx\ \&\ Rxcd)$

P1. $a \leq a$

P2. $(a \leq b\ \&\ Rbcd) \Rightarrow Racd$

(3) a (valuation) relation $\vDash$ from K to the set of all wffs $\mathcal{F}$ such that the following conditions (clauses) are satisfied for every $p \in \mathcal{P}$, $A, B \in \mathcal{F}$, and

$a \in K$:

(i). $(a \leq b \ \& \ a \vDash p) \Rightarrow b \vDash p$

(ii). $a \vDash A \wedge B$ iff $a \vDash A$ and $a \vDash B$

(iii). $a \vDash A \vee B$ iff $a \vDash A$ or $a \vDash B$

(iv). $a \vDash A \rightarrow B$ iff for all $b, c \in K$, $(Rabc \ \& \ b \vDash A) \Rightarrow c \vDash B$

Additional elements of M are the following: (1) a set of postulates $\mathrm{Pj}_1, ..., \mathrm{Pj}_n$; (2) a subset S of K such that $O \cap S \neq \emptyset$, and (3) the following conditions (clauses) for every $a \in K$:

(v). $a \vDash \neg A$ iff for all $b, c \in K$, $(Rabc \ \& \ c \in S) \Rightarrow b \nvDash A$

(vi). $a \vDash \neg A$ iff for all $b, c \in K$, $Rabc \Rightarrow b \nvDash A$

(vii). $a \vDash f$ iff $a \notin S$

(viii). $a \vDash t$ iff $a \in O$

(ix). $a \nvDash f$

If clause (vii) (respectively, (viii)) is added, then the postulate Pf (respectively, Pt) noted below is also added for all $a, b \in K$:

$$Pf. \ (a \leq b \ \& \ b \in S) \Rightarrow a \in S$$
$$Pt. \ (a \leq b \ \& \ a \in O) \Rightarrow b \in O$$

Structures of the form $(K, O, R, \vDash)$ satisfying d1, d2, P1, P2 and clauses (i)–(iv) are the basic structures and in fact characterize the logic B₊ (they are labelled "B₊-models"; cf. Definition 1.12). Introduction of additional elements recorded above serve to determine extensions or expansions of B₊ interpretable in models with a set of designated points O. Next, the notions of truth and validity are defined.

Definition 1.5 (Truth in a class of EB₊-models). Let a class of EB₊-models $\mathcal{M}$ be defined and $A \in \mathcal{F}$. Then, A is true in a model $M \in \mathcal{M}$ iff $a \vDash A$ for all $a \in O$ in M. (Given a model M, $a \nvDash A$ symbolizes that A is not true in M.)

Definition 1.6 (Validity in a class of EB₊-models). Let a class of EB₊-models $\mathcal{M}$ be defined and $A \in \mathcal{F}$. Then, A is $\mathcal{M}$-valid (in symbols $\vDash_{\mathcal{M}} A$) iff A is true in all $M \in \mathcal{M}$. ($\nvDash_{\mathcal{M}} A$ symbolizes that A is not $\mathcal{M}$-valid.)

Subscripts are generally deleted when there is no danger of confusion. By the term EB_+-validity we shall generally refer to validity in any class of EB_+-models, $\mathcal{M}$.

As pointed out in the Introduction, we shall understand that the definition of a class of models together with the definition of validity, given that class of models, determine a semantics that will be referred to by the name assigned to the said class of models.

Definition 1.7 (RM-semantics for EB_+-logics). Let L be an EB_+-logic. Σ is a semantics for L iff L is sound and complete w.r.t. Σ. In this sense, we say that L-models (together with the definition of L-validity) constitute an RM-semantics for L. For example, below it is shown that B_+-semantics is an RM-semantics for the logic B_+.

We prove a couple of lemmas that will be used in the soundness proofs throughout this book.

By d1, a binary relation is introduced in ternary models. The following useful lemma shows that truth is hereditary w.r.t. this binary relation: if $a \leq b$ all formulas which are true in a are also true in b. (Definition d2 is used in some presentations of B_+ —cf., e.g., [35]—. Here it is only employed for defining extensions and expansions of B_+ —cf. Chapter 6.)

Lemma 1.8 (Hereditary Condition). *For any EB_+-model, $a, b \in K$ and $A \in \mathcal{F}$, $(a \leq b \,\&\, a \vDash A) \Rightarrow b \vDash A$.*

Proof. (In this and the rest of the proofs to follow, by i, ii, etc., we refer to clauses (i), (ii), etc. in Definition 1.4.) Given an arbitrary EB_+-model, the proof is by induction on the length of A. (a) A is a propositional variable: by i. (b) A is $B \wedge C$ or $B \vee C$: the proof is trivial by ii and iii, respectively. (c) A is $B \to C$: suppose (1) $a \leq b$, (2) $a \vDash B \to C$, (3) $Rbcd$ for $c, d \in K$, and (4) $c \vDash B$. We have to prove $d \vDash C$. By 1, 3, and P2, we have (5) $Racd$, whence (6) $d \vDash C$ follows by 2, 4, 5, and iv. (d) A is $\neg B$: suppose (1) $a \leq b$, (2) $a \vDash \neg B$, (3) $Rbcd$ for $c, d \in K$, and (4) $d \in S$. We have to prove $c \nvDash B$. By 1, 3, and P2, we have (5) $Racd$, whence (6) $c \nvDash B$ follows by 2, 4, and v. (If clause (vi) has been introduced instead of clause (v), the proof is similar: actually, it suffices to delete any reference to the set S.) (e) A is f: Suppose (1) $a \leq b$ and (2) $a \vDash f$. By vii, we have (3) $a \notin S$. By 1, 3, and Pf, we have (4) $b \notin S$, whence (5) $b \vDash f$ follows, as it was required. (If clause (ix) has been introduced instead of clause (vii), $(a \leq b \,\&\, a \vDash f) \Rightarrow b \vDash f$ is immediate.) (f) A is t: the proof is immediate by viii and Pt. $\qquad \square$

In addition to being useful in the proof of soundness, the following lemma, the Entailment Lemma, is significant in itself: a conditional is valid iff, in any model, the consequent is true in a given world whenever the antecedent is true in that same world.

Lemma 1.9 (Entailment Lemma). *Let a class of EB$_+$-models $\mathcal{M}$ be defined. For any $A, B \in \mathcal{F}$, $\vDash_{\mathcal{M}} A \to B$ iff $(a \vDash A \Rightarrow a \vDash B$ for all $a \in K)$ in all $M \in \mathcal{M}$.*

Proof. Let $M \in \mathcal{M}$. From left to right (a) (($\Rightarrow$) henceforward). We have (1) For any $a \in K$, $Rxaa$ for some $x \in O$ (P1, d1) and (2) $x \vDash A \to B$, by Definition 1.6, given $\vDash_{\mathcal{M}} A \to B$. Then, by 1 and 2 we have (3) if $a \vDash A$, then $a \vDash B$ by iv. From right to left (b) (($\Leftarrow$) henceforward). Suppose (1) $Rabc$ $(a \in O)$ and (2) $b \vDash A$. We have to prove $c \vDash B$. By 2 and the hypothesis of the case, we get (3) $b \vDash B$. By 1 and d1, we get (4) $b \leq c$. Finally, (5) $c \vDash B$ follows by 3, 4, and Lemma 1.8. $\square$

1.2 THE LOGIC B$_+$

Next, the basic positive logic in RM$_1$-semantics (RM-semantics with a set of designated points), the logic B$_+$, is defined and some of its theorems, which are useful in the completeness proof, are remarked.

Definition 1.10 (The logic B$_+$). The logic B$_+$ is axiomatized with the following axioms and rules of inference:

Axioms

$\quad$ A1. $A \to A$

$\quad$ A2. $(A \wedge B) \to A$ / $(A \wedge B) \to B$

$\quad$ A3. $[(A \to B) \wedge (A \to C)] \to [A \to (B \wedge C)]$

$\quad$ A4. $A \to (A \vee B)$ / $B \to (A \vee B)$

$\quad$ A5. $[(A \to C) \wedge (B \to C)] \to [(A \vee B) \to C]$

$\quad$ A6. $[A \wedge (B \vee C)] \to [(A \wedge B) \vee (A \wedge C)]$

Rules

$\quad$ Modus ponens (MP). A & $A \to B \Rightarrow B$

$\quad$ Adjunction (Adj). A & $B \Rightarrow A \wedge B$

$\quad$ Suffixing (Suf). $A \to B \Rightarrow (B \to C) \to (A \to C)$

$\quad$ Prefixing (Pref). $B \to C \Rightarrow (A \to B) \to (A \to C)$

Proposition 1.11 (Some theorems of B_+). *The following rules and wffs, which are provable in B_+, will be useful in the completeness proof:*

Transitivity (Trans). $A \to B$ & $B \to C \Rightarrow A \to C$

Introduction of conjunction $(I \wedge_1)$. $A \to B$ & $A \to C \Rightarrow A \to (B \wedge C)$

Introduction of conjunction $(I \wedge_2)$. $A \to C$ & $B \to D \Rightarrow (A \wedge B) \to (C \wedge D)$

Elimination of disjunction $(E \vee_1)$. $A \to C$ & $B \to C \Rightarrow (A \vee B) \to C$

Elimination of disjunction $(E \vee_2)$. $A \to C$ & $B \to D \Rightarrow (A \vee B) \to (C \vee D)$

t1. $[A \wedge (B \wedge C)] \to [(A \wedge B) \wedge (A \wedge C)]$

t2. $[(A \vee B) \wedge (C \wedge D)] \to [(A \wedge C) \vee (B \wedge D)]$

t3. $[(A \to C) \vee (B \to D)] \to [(A \wedge B) \to (C \vee D)]$

t4. $[(A \to C) \wedge (B \to D)] \to [(A \vee B) \to (C \vee D)]$

t5. $[(A \to C) \wedge (B \to D)] \to [(A \wedge B) \to (C \wedge D)]$

t6. $[(A \vee B) \to C] \leftrightarrow [(A \to C) \wedge (B \to C)]$

Proof. Cf. [37], Chapter 4; [24], Chapter 1. We note that Trans, $I\wedge_1$, $I\wedge_2$, $E\vee_1$, $E\vee_2$, t1, and t2 are provable in the positive fragment, FDE_+, of Anderson and Belnap's First Degree Entailment Logic, FDE (cf. [2], [37]). (The logic FDE_+ is axiomatized with A1, A2, A4, A6, MP, Adj, $E\vee_1$, and $I\wedge_1$; cf. [39].) □

The basic models among EB_+-models are B_+-models, which determine the logic B_+ just defined.

Definition 1.12 (B_+-models). A B_+-model is a basic EB_+-model with no additional items added. That is, a B_+-model is a structure $(K, O, R, \vDash)$ where R is subject to d1, d2, P1, and P2, and $\vDash$, to just clauses (i)–(iv) (we recall that validity in an EB_+-model —and so, B_+-validity— is understood according to Definition 1.6. Cf. also Definitions 1.4, 1.5).

In what follows, we will proceed to the proof of the soundness and completeness of B_+ in a general way so that the results obtained can be used in the soundness and completeness proofs of the extensions and expansions of B_+ (including extensions and expansions of B_{K+}) treated in the present work.

Soundness of B_+ (cf. Corollary 1.29) follows from the general proposition proved below. (Recall that EB_+-valid means valid in any class of EB_+-models; cf. Definition 1.6.)

Proposition 1.13 (All theorems of B_+ are EB_+-valid). *For any $A \in \mathcal{F}$, if $\vdash_{B_+} A$, then A is EB_+-valid.*

Proof. The proof is simplified by using Lemma 1.9.

I. *The axioms are EB_+-valid*: A1, A2, A4, and A6 are immediate. So, let us prove A3 and A5 (we prove A5: the proof of A3 is similar).

A5, $[(A \to C) \wedge (B \to C)] \to [(A \vee B) \to C]$, is EB_+-valid: Let $a \in K$ in an arbitrary EB_+-model M and suppose, for $A, B, C \in \mathcal{F}$, (1) $a \vDash (A \to C) \wedge (B \to C)$. We have to prove $a \vDash (A \vee B) \to C$. So, suppose for $b, c \in K$ in M, (2) $Rabc$ and (3) $b \vDash A \vee B$. Then, A5 is true in M if we prove $c \vDash C$. Firstly, (by 1 and ii) we have (4) $a \vDash A \to C$ and (5) $a \vDash B \to C$. Suppose now the first alternative in 3, (6) $b \vDash A$. We get (7) $c \vDash C$ (by 2, 4, 6, and iv). On the other hand, suppose the second alternative in 3, (8) $b \vDash B$. We have again (9) $c \vDash C$ (by 2, 5, 8, and iv). Thus, $c \vDash C$ is proved, as it was required.

II. *The rules preserve EB_+-validity*:

(a) MP, $A \& A \to B \Rightarrow B$, preserves EB_+-validity: Let $a \in O$ in an arbitrary EB_+-model M and suppose for $A, B \in \mathcal{F}$, (1) $\vDash A$ and (2) $\vDash A \to B$. We need to prove $a \vDash B$. By d1 and P1, we have (3) $Rxaa$ for some $x \in O$ in M. Then, we have (4) $a \vDash A$ (by 1 and Definition 1.6) and (5) $x \vDash A \to B$ (by 1 and Definition 1.6, since $x \in O$). Finally, (6) $a \vDash B$ follows (by 3, 4, 5, and iv), as it was to be proved.

(b) Adj, $A \& B \Rightarrow A \wedge B$, preserves EB_+-validity. The proof is trivial.

(c) Suf, $A \to B \Rightarrow (B \to C) \to (A \to C)$, preserves EB_+-validity: Let $a \in K$ in an arbitrary EB_+-model M and suppose for $A, B, C \in \mathcal{F}$, (1) $\vDash A \to B$ and (2) $a \vDash B \to C$. We have to prove $a \vDash A \to C$. So, suppose, for $b, c \in K$ in M, (3) $Rabc$ and (4) $b \vDash A$. We now need to prove $c \vDash C$. We have (5) $b \vDash B$ (by 1, 4, and Lemma 1.9) and then (6) $c \vDash C$ (by 2, 3, 5, and iv).

(d) Pref, $B \to C \Rightarrow (A \to B) \to (A \to C)$, preserves EB_+-validity: The proof is similar to that of Suf. □

1.3 COMPLETENESS OF EB₊-LOGICS I. BASIC PROPOSITIONS AND LEMMAS

In the sequel, we establish the basic facts for proving the completeness of B_+ and that of its extensions and its expansions to be defined. (Recall that by an EB_+-logic, we generally mean an extension or expansion of B_+. We essentially follow the ideas, strategy, and terminology of [37], Chapter 4.)

Next, the notion of a theory and the classes of theories which are of interest in the present work are defined.

Definition 1.14 (EB$_+$-theories). Let L be an EB$_+$-logic. An L-theory is a set of formulas closed under Adjunction (Adj) and L-implication (L-imp). That is, a is an L-theory if, for $A, B \in \mathcal{F}$, whenever $A, B \in a$, then $A \wedge B \in a$; and if, for $A, B \in \mathcal{F}$, whenever $A \rightarrow B$ is a theorem of L and $A \in a$, then $B \in a$.

Definition 1.15 (Classes of EB$_+$-theories). Let L be an EB$_+$-logic and a be an L-theory. We set: (1) a is prime iff, for $A, B \in \mathcal{F}$, whenever $A \vee B \in a$, then $A \in a$ or $B \in a$. (2) a is non-empty iff it contains at least a wff. (3) a is trivial iff every wff belongs to it. (4) a is regular iff it contains all theorems of L. (5) a is weak-consistent (consistent in a weak sense) iff it is not w-inconsistent (a is w-inconsistent —inconsistent in a weak sense— iff it contains the negation of a theorem of L). (6) a is u-consistent (weakly consistent in a second sense) iff it is not u-inconsistent (a is u-inconsistent —weakly inconsistent in a second sense— iff $A \in a$, $\neg A$ being a theorem of L). (7) a is a-consistent (consistent in an absolute sense) iff a is not trivial.

The completeness proofs in the present work are based upon canonical model constructions. The following definition introduces the fundamental notions for defining these canonical models. Then, a series of lemmas follow establishing the basic facts upon which the completeness proofs to be developed in the sequel are based. These basic facts hold for any EB$_+$-logic L. In Definition 1.27, the canonical B$_+$-model is defined and in Theorem 1.30, the completeness of B$_+$ is demonstrated.

Definition 1.16 (Pre-canonical notions). Let L be an EB$_+$-logic and K^T be the set of all L-theories. Then, the ternary relation R^T is defined on K^T as follows: for all $A, B \in \mathcal{F}$ and $a, b, c \in K^T$, $R^T abc$ iff ($A \rightarrow B \in a$ & $A \in b$) $\Rightarrow B \in c$. Now, let K^P be the set of all prime L-theories. Then, the ternary relation R^P is the restriction of R^T to K^P. On the other hand, O^T is the set of all regular L-theories, U^T is the set of all u-consistent L-theories, and S^T is the set of all w-consistent L-theories, whereas O^P, U^P, and S^P are the subsets of O^T, U^T, and S^T in K^P, respectively. In addition, the superscripts N and W abbreviate "non-empty" and "w-consistent". Thus, for example, K^{NT} is the set of all non-empty L-theories; and O^{WT} is the set of all w-consistent and regular L-theories. Finally, the relation $\vDash^P$ is defined as follows: for all $A \in \mathcal{F}$ and $a \in K^P$, $a \vDash^P A$ iff $A \in a$. (Notice that the relation $\vDash^P$ can be labelled by the superscripts N and/or W.)

The ensuing lemma shows how to extend non-trivial theories to prime ones.

Lemma 1.17 (Primeness lemma). *Let $a \in K^T$ and A be a wff such that $A \notin a$. Then, there is $x \in K^P$ such that $a \subseteq x$ and $A \notin x$.*

Proof. By using, for example, Kurakowski-Zorn's Lemma, a is extended to a maximal L-theory x such that $a \subseteq x$ and $A \notin x$. If x is not prime, then there are $B, C \in \mathcal{F}$ such that $B \vee C \in x$ but $B \notin x$ and $C \notin x$. We define the set $[x, B] = \{E \mid \exists D[D \in x \ \& \vdash_L (B \wedge D) \to E]\}$. The set $[x, C]$ is defined similarly. Below we prove the following: (I) $[x, B]$ and $[x, C]$ are L-theories; (II) $x \subset [x, B]$, $x \subset [x, C]$. Given (I) and (II), by the maximality of x, we have (III) $A \in [x, B]$ and $A \in [x, C]$, whence it will be easy to prove (IV) $A \in x$, a contradiction. Therefore, x is prime. (Since there is no danger of confusion, the subscript L is omitted under $\vdash$ in this proof and the ones to follow. On the other hand, facts (I) and (II) are proved for $[x, B]$, the proof for $[x, C]$ being similar.)

(I). $[x, B]$ and $[x, C]$ are L-theories. We prove that $[x, B]$ is an L-theory:

(a) $[x, B]$ is closed under L-imp: Suppose for $D, E \in \mathcal{F}$, (1) $\vdash D \to E$ and (2) $D \in [x, B]$. We have (3) $\vdash (B \wedge F) \to D$ for some $F \in x$ (2, definition of $[x, B]$), (4) $\vdash (B \wedge F) \to E$ (1, 3, Trans) and, finally, (5) $E \in [x, B]$ (4, definition of $[x, B]$, since $F \in x$ by 3).

(b) $[x, B]$ is closed under Adj: Suppose, for $D, E \in \mathcal{F}$, (1) $D \in [x, B]$ and (2) $E \in [x, B]$. By definition of $[x, B]$ we have (3) $\vdash (B \wedge F) \to D$ ($F \in x$) and (4) $\vdash (B \wedge F') \to E$ ($F' \in x$). Then, (5) $\vdash [(B \wedge F) \wedge (B \wedge F')] \to (D \wedge E)$ follows (by 3, 4, and the rule I$\wedge_2$) whence we get (6) $\vdash [B \wedge (F \wedge F')] \to (D \wedge E)$ by t1, $[B \wedge (F \wedge F')] \to [(B \wedge F) \wedge (B \wedge F')]$, and Trans. Finally, we have (7) $D \wedge E \in [x, B]$ (by 6 and definition of $[x, B]$ since $F \wedge F' \in x$ —by 3 and 4 as x is an L-theory and consequently it is closed under Adj).

(II). $x \subset [x, B]$ and $x \subset [x, C]$. We prove $x \subset [x, B]$:

(a) $x \subseteq [x, B]$: Suppose $D \in x$. By A2, we have $\vdash (B \wedge D) \to D$. Thus, $D \in [x, B]$ follows.

(b) $x \subset [x, B]$: Let $D \in x$ (notice that x is non-empty since, by hypothesis, $B \vee C \in x$). By A2, we have $\vdash (B \wedge D) \to B$. Thus, $B \in [x, B]$ follows. But by hypothesis $B \notin x$.

(IV). Given I and II we get (III) $A \in [x, B]$ and $A \in [x, C]$ by the maximality of x. Then, we have (1) $\vdash (B \wedge D) \to A$ ($D \in x$) and (2) $\vdash (C \wedge D') \to A$ ($D' \in x$). By the rule E$\vee_1$, (3) $\vdash [(B \wedge D) \vee (C \wedge D')] \to A$ follows, whence by t2, $[(B \vee C) \wedge (D \wedge D')] \to [(B \wedge D) \vee (C \wedge D')]$, and Trans, we get (4) $\vdash [(B \vee C) \wedge (D \wedge D')] \to A$. Now, we have (5) $(B \vee C) \wedge (D \wedge D') \in x$ since $B \vee C \in x$ by hypothesis and $D \wedge D' \in x$ by 1 and 2 (x is closed under Adj). Then, we have (6) $A \in x$ by 4 and 5.

Therefore x is prime. □

We remark that the proof of Lemma 1.17 can be carried out in the context of logics weaker than B_+. Actually, the positive fragment of Anderson and Belnap's First Degree Entailment logic is sufficient (cf. Proposition 1.11 on the logic FDE_+).

The lemma that follows shows how to extend the first member in a ternary relation to a prime theory.

Lemma 1.18 (Extending a in $R^T abc$ to a member in K^P). *Let $a, b \in K^T$, $c \in K^P$, and $R^T abc$. Then, there is some $x \in K^P$ such that $a \subseteq x$ and $R^T xbc$.*

Proof. We extend a to a maximal L-theory x such that $a \subseteq x$ and $R^T xbc$. If x is not prime, then there are wffs B, C such that $B \vee C \in x$ but $B \notin x$ and $C \notin x$. We define the sets $[x, B]$ and $[x, C]$ similarly as in Lemma 1.17. Correspondingly as in this lemma, $[x, B]$ and $[x, C]$ are proved L-theories strictly including x. By the maximality of x, we have (1) not-$R^T [x, B]bc$ and (2) not-$R^T [x, C]bc$. But below it is shown that a contradiction follows from 1 and 2, whence it has to be concluded that x is prime.

By 1, 2, and definition of R^T, we have for $A, A', D, D' \in \mathcal{F}$ (3) $A \rightarrow D \in [x, B]$, (4) $A' \rightarrow D' \in [x, C]$, (5) $A \in b$ and $A' \in b$, and (6) $D \notin c$ and $D' \notin c$. Moreover, we have (7) $\vdash (B \wedge E) \rightarrow (A \rightarrow D)$ ($E \in x$) and (8) $\vdash (C \wedge E') \rightarrow (A' \rightarrow D')$ ($E' \in x$) by applying the definitions of $[x, B]$ and $[x, C]$ to 3 and 4, respectively. Then, we proceed as follows. By 7, 8, A4, and Trans, we get (9) $\vdash (B \wedge E) \rightarrow [(A \rightarrow D) \vee (A' \rightarrow D')]$ and (10) $\vdash (C \wedge E') \rightarrow [(A \rightarrow D) \vee (A' \rightarrow D')]$ whence, by the rule $E\vee_1$, we have (11) $\vdash [(B \wedge E) \vee (C \wedge E')] \rightarrow [(A \rightarrow D) \vee (A' \rightarrow D')]$ and by t2, $[(B \vee C) \wedge (E \wedge E')] \rightarrow [(B \wedge E) \vee (C \wedge E')]$, and Trans, (12) $\vdash [(B \vee C) \wedge (E \wedge E')] \rightarrow [(A \rightarrow D) \vee (A' \rightarrow D')]$. Given $(B \vee C) \wedge (E \wedge E') \in x$ (by the hypothesis, 7, and 8, since x is closed under Adj), we have (13) $(A \rightarrow D) \vee (A' \rightarrow D') \in x$ by 12. Next, we conclude the following. By t3, (14) $\vdash [(A \rightarrow D) \vee (A' \rightarrow D')] \rightarrow [(A \wedge A') \rightarrow (D \vee D')]$; by 13 and 14, (15) $(A \wedge A') \rightarrow (D \vee D') \in x$. Finally, by $R^T xbc$, 5, 15, and definition of R^T, (16) $D \vee D' \in x$. But 6 and 16 contradict each other. $\square$

On the other hand, Lemma 1.19 shows how the second member in a given ternary relation can be extended to a prime theory.

Lemma 1.19 (Extending b in $R^T abc$ to a member in K^P). *Let $a, b \in K^T$, $c \in K^P$ and $R^T abc$. Then, there is some $x \in K^P$ such that $b \subseteq x$ and $R^T axc$.*

Proof. We extend b to a maximal L-theory x such that $b \subseteq x$ and $R^T axc$. If, by reductio, x is not prime, then there are wffs B, C such that $B \vee C \in x$ but $B \notin x$ and $C \notin x$. Just as in Lemmas 1.17 and 1.18, it is proved that the sets $[x, B]$ and $[x, C]$ are L-theories strictly including x. By maximality of x, (1) not-$R^T a[x, B]c$ and not-$R^T a[x, C]c$. Similarly as in Lemma 1.18, it is shown below that a contradiction follows from 1, whence x is prime.

By 1 and definition of R^T, we have for $A, A', D, D' \in \mathcal{F}$ (2) $A \to D \in a$ and $A' \to D' \in a$, (3) $A \in [x, B]$ and $A' \in [x, C]$ and (4) $D \notin c$ and $D' \notin c$. And by 3 and definition of $[x, B]$ and $[x, C]$ we get (5) $\vdash (B \wedge E) \to A$ and $\vdash (C \wedge E') \to A'$ with $E \in x$, $E' \in x$. By 5, A4, and Trans, we infer (6) $\vdash (B \wedge E) \to (A \vee A')$ and $\vdash (C \wedge E') \to (A \vee A')$ whence, by the rule E$\vee_1$, we get (7) $\vdash [(B \wedge E) \vee (C \wedge E')] \to (A \vee A')$, and by t2, $[(B \vee C) \wedge (E \wedge E')] \to [(B \wedge E) \vee (C \wedge E')]$, and Trans, (8) $\vdash [(B \vee C) \wedge (E \wedge E')] \to (A \vee A')$. Now, we have (9) $A \vee A' \in x$ by 8 and closure of x under Adj and under L-imp, since $(B \vee C) \wedge (E \wedge E') \in x$ ($B \vee C \in x$ —by the hypothesis— and $E \wedge E' \in x$ by 5). On the other hand, by t4 we have (10) $\vdash [(A \to D) \wedge (A' \to D')] \to [(A \vee A') \to (D \vee D')]$. Then, by 2 and 10, we derive (11) $(A \vee A') \to (D \vee D') \in a$ and, finally, by 9, 11, and $R^T axc$, we conclude (12) $D \vee D' \in c$. But 4 and 12 contradict each other. $\square$

Consider the following definitions.

Definition 1.20 ($a \leq^T b$). Let $a, b \in K^T$. $a \leq^T b =_{df} \exists x \in O^T Rxab$.

Definition 1.21 ($a \leq^P b$). Let $a, b \in K^P$. $a \leq^P b =_{df} \exists x \in O^P Rxab$.

Lemma 1.22 proves that the binary relation introduced via definition d1 (cf. Definition 1.4) is nothing but set inclusion between L-theories from a pre-canonical point of view.

Lemma 1.22 ($a \leq^T b$ iff $a \subseteq b$). *For any* $a, b \in K^T$, $a \leq^T b$ *iff* $a \subseteq b$.

Proof. Let $a, b \in K^T$. (a) ($\Rightarrow$) Suppose $a \leq^T b$. Then, we have $R^T xab$ for some $x \in O^T$. Moreover, suppose $A \in a$ for $A \in \mathcal{F}$. By A1, $A \to A \in x$ follows. Then, we have $A \in b$. (b) ($\Leftarrow$) Suppose $a \subseteq b$. As a is closed under L-imp, we get $R^T Laa$. Clearly, we have $R^T Lab$, for $a \subseteq b$. Thus, $a \leq^T b$ follows, since $L \in O^T$. $\square$

Corollary 1.23 ($a \leq^P b$ iff $a \subseteq b$). *For any* $a, b \in K^P$, $a \leq^P b$ *iff* $a \subseteq b$.

Proof. Let $a, b \in K^P$. (a) ($\Rightarrow$) Suppose $a \leq^P b$. By Lemma 1.22, we have $a \subseteq b$. (b) ($\Leftarrow$) Suppose $a \subseteq b$. By Lemma 1.22, there is some $y \in O^T$ such that $R^T yab$. By Lemma 1.18, there is some $x \in O^P$ such that $y \subseteq x$ and $R^P xab$. Thus, we have $a \leq^P b$, as it was required. $\square$

The lemma that follows indicates how to build a L-theory x such that $R^T abx$ once we are given the L-theories a and b.

Lemma 1.24 (Defining x for a, b in R^T). *Let $a, b \in K^T$. The set $x = \{B \mid \exists A[A \rightarrow B \in a \ \& \ A \in b]\}$ is an L-theory such that $R^T abx$.*

Proof. Firstly, we prove that x is an L-theory.

(a) x is closed under L-imp. Suppose for $A, B \in \mathcal{F}$ (1) $\vdash A \rightarrow B$ and (2) $A \in x$. By definition of x, we have (3) there is some $C \in b$ such that $C \rightarrow A \in a$. By 1 and Pref, we get (4) $\vdash (C \rightarrow A) \rightarrow (C \rightarrow B)$, whence by 3, we infer (5) $C \rightarrow B \in a$, i.e., $B \in x$, as $C \in b$.

(b) x is closed under Adj. Suppose (1) $A, B \in x$. By definition of x we have (2) $C \rightarrow A \in a$ and $C' \rightarrow B \in a$ for some $C, C' \in b$, that is, (3) $(C \rightarrow A) \wedge (C' \rightarrow B) \in a$ by closure of a under Adj. By t5, we get (4) $\vdash [(C \rightarrow A) \wedge (C' \rightarrow B)] \rightarrow [(C \wedge C') \rightarrow (A \wedge B)]$. By 3 and 4, we deduce (5) $(C \wedge C') \rightarrow (A \wedge B) \in a$. Now, (6) $C \wedge C' \in b$, by 2. So, (7) $A \wedge B \in x$, by 5 and 6, as it was required.

Finally, $R^T abx$ follows immediately by definitions of x and R^T. ☐

The next, and last important lemma, demonstrates that the pre-canonical valuation relation $\vDash^P$ is in fact a valuation relation (cf. Definition 1.4) when it is understood as membership of formulas in prime L-theories. (Notice that we restrict ourselves to the minimal positive language with the connectives $\rightarrow$, $\wedge$, and $\vee$. The remaining connectives —$\neg, f$, and t— are treated in subsequent chapters.)

Lemma 1.25 (On the pre-canonical relation $\vDash^P$). *For any $p \in \mathcal{P}$, $A, B \in \mathcal{F}$, and $a, b \in K^P$, we have*

(a) $(a \subseteq b \ \& \ p \in a) \Rightarrow p \in b$

(b) $A \wedge B \in a$ iff $A \in a$ and $B \in a$

(c) $A \vee B \in a$ iff $A \in a$ or $B \in a$

(d) If $A \rightarrow B \in a$, then for all $b, c \in K^P$, $(R^P abc \ \& \ A \in b) \Rightarrow B \in c$

(e) If $A \rightarrow B \notin a$, then there are $x, y \in K^P$ such that $R^P axy$, $A \in x$, and $B \notin y$

Proof. (a) It is obvious. (b) ($\Rightarrow$) by A2 and closure of a under L-imp; (b) ($\Leftarrow$) by closure of a under Adj. (c) ($\Rightarrow$) by primeness of a; (c) ($\Leftarrow$) by A4 and closure of a under L-imp. (d) It is immediate by definition of R^P. (e) Suppose there are $A, B \in \mathcal{F}$ and $a \in K^P$ such that $A \rightarrow B \notin a$. Consider the sets

$z = \{C \mid\vdash_L A \rightarrow C\}$ and $u = \{C \mid \exists D[D \rightarrow C \in a \ \& \ D \in z]\}$. It is easy to prove that z and u are L-theories (cf. the proofs that $[x, B]$ and x are L-theories in Lemmas 1.17 and 1.24, respectively). Then, $R^T azu$ is derived by Lemma 1.24, $A \in z$, by A1 and $B \notin u$ is proved as follows: if $B \in u$, then for some $C \in \mathcal{F}$, $C \rightarrow B \in a$, and $\vdash_L A \rightarrow C$ whence by Suf $A \rightarrow B \in a$ contradicting the hypothesis. Consequently, we have L-theories z and u such that $R^T azu$, $A \in z$, and $B \notin u$. Now, by Lemma 1.17, there is some $y \in K^P$ such that $u \subseteq y$ and $B \notin y$. Obviously, $R^T azy$. Next, by Lemma 1.19, there is some $x \in K^P$ such that $z \subseteq x$ and $R^P axy$. Therefore, we have prime L-theories x, y such that $A \in x$, $B \notin y$, and $R^P axy$, as was to be proved.   □

1.4 COMPLETENESS OF EB$_+$-LOGICS II. CANONICAL MODELS. COMPLETENESS OF B$_+$

In this section, we present a proof of the soundness and completeness of B$_+$. As it was pointed out above, the completeness proofs in the present work are based upon canonical model constructions. In Definition 1.26, the notion of a canonical EB$_+$-model is defined in a general way. The canonical B$_+$-model is a particular EB$_+$-model.

In the same sense that B$_+$models are the basic EB$_+$-models, canonical B$_+$models are the simplest canonical EB$_+$-models. It has undoubtedly been remarked that, so far, L-theories have not been required to have any special properties except primeness. However, non-emptiness and/or w-consistency can be essential in the case of some EB$_+$-logics interpretable with RM-semantics. Therefore, the following types of canonical EB$_+$-models will be considered in the course of the present book.

Definition 1.26 (Canonical EB$_+$-models). There are four possible types of canonical EB$_+$-models for EB$_+$-logics: (a) $(K^P, O^P, R^P, \vDash^P)$; (b) $(K^P, O^P, S^P, R^P, \vDash^P)$; (c) $(K^{NP}, O^P, S^{NP}, R^{NP}, \vDash^{NP})$; and (d) $(K^{WNP}, O^{WP}, R^{WNP}, \vDash^{WNP})$ where the relation $\vDash^P$, the sets K, O, and S and superscripts P, N, and W are read as in Definition 1.16.

Thus, for instance, all L-theories are non-empty and prime in canonical EB$_+$-models of type (c); and all L-theories are w-consistent, non-empty and prime in canonical EB$_+$-models of type (d). On the other hand, $\vDash^{WNP}$ is the (valuation) relation between the set K^{WNP} and the set $\mathcal{F}$ of the language of L in question.

We note that canonical models with u–consistent L–theories are only marginally used in the sequel (cf. Chapters 4 and 6). Consequently, they are omitted in Definition 1.26, for simplicity reasons.

Definition 1.27 (The canonical B_+-model). The canonical B_+-model is a canonical EB_+-model of type a (cf. Definition 1.26). That is, it is the structure $(K^P, O^P, R^P, \vDash^P)$. (Recall that $K^P, O^P, R^P, \vDash^P$ are here referred to B_+-theories; cf. Definition 1.16.)

The next Lemma shows that the canonical B_+-model is indeed a B_+-model.

Lemma 1.28 (The canonical B_+-model is a B_+-model). *The canonical B_+-model is indeed a B_+-model.*

Proof. Given that R^P is clearly a ternary relation on K^P and that $\vDash^P$ is obviously a relation from K^P to the set of all wffs $\mathcal{F}$, it remains to prove the facts (1)–(3) listed below.

(1) The set O^P is not empty. The logic B_+ is, of course, a regular, non-trivial B_+-theory. By Lemma 1.17, there is a regular and prime L-theory x such that $B_+ \subseteq x$.

(2) Postulates P1 and P2 hold in the canonical B_+-model: P1 is immediate by Corollary 1.23 and P2 is immediate by Corollary 1.23 and definition of R^P.

(3) Clauses (i)–(iv) hold in the canonical B_+-model. The proof is immediate by Lemma 1.25 and definition of $\vDash^P$. Let us consider, for example, clause (ii). By Lemma 1.25, $A \wedge B \in a$ iff $A \in a$ and $B \in a$, for any $a \in K^P$. Thus, $a \vDash^P A \wedge B$ iff $a \vDash^P A$ and $a \vDash^P B$ by definition of $\vDash^P$.   □

Finally, we prove (weak) soundness and completeness of B_+ w.r.t. the semantics enclosed in Definitions 1.12, 1.5, and 1.6.

Corollary 1.29 (Soundness of B_+). *For any $A \in \mathcal{F}$, if $\vdash_{B_+} A$, then $\vDash_{B_+} A$.*

Proof. Immediate by Proposition 1.13, since a B_+-model is an EB_+-model.   □

Theorem 1.30 (Completeness of B_+). *For any $A \in \mathcal{F}$, if $\vDash_{B_+} A$, then $\vdash_{B_+} A$.*

Proof. Suppose $\nvdash_{B_+} A$. By Lemma 1.17, there is a regular, prime L-theory x such that $B_+ \subseteq x$ and $A \notin x$. By Definition 1.27 and Lemma 1.28, $x \nvDash^C A$. Therefore, $\nvDash_{B_+} A$ by Definition 1.12.   □

We close the section with the following remark.

Remark 1.31 (On the proof of the completeness theorems). Let L be any EB₊-logic. The proof that L is complete follows the pattern set in Theorem 1.30. Firstly, the canonical L-model is shown an L-model according to the method procured by Lemma 1.28. That is, after proving that the canonical set (K^P, K^{NP}, or K^{WNP}, as the case may be —cf. Definition 1.26—) is not empty, it is proved that the semantical postulates and clauses hold in the canonical model. Then, completeness follows by demonstrating that every non-theorem A fails to belong to a member a of the canonical set, whence A is not true in a and consequently is not valid in the RM-semantics for L.

1.5 "RULES OF INFERENCE", "RULES OF PROOF", AND STRONG COMPLETENESS

A rule r of a logic L is a *rule of inference* if it can be applied to no matter which premises formulated in the language of L; and r is a *rule of proof* if it is applied only to theorems of L. The distinction is essential in some substructural logics formulated in the Hilbert-style way. The case of relevance logics is paradigmatic. Actually, to the best of our knowledge, Ackermann (the father of relevance logics) was the first logician who defined a logic (his systems Π and Π', cf. [1]) whose formulation leans essentially in the aforementioned distinction. Although he did not define the notions involved, Ackermann strongly stressed (cf. [1], pp. 119–120) that the rule δ (i.e., $B \,\&\, A \to (B \to C) \Rightarrow A \to C$) can only be applied if B is a "logische Identität" (a logical theorem). A number of relevance logics are formulated with both rules of inference and rules of proof. For example, Anderson and Belnap's Logic of Entailment E (cf. [2]), Routley and Meyer's basic logic B (cf. [37]) or Brady's weak relevance logics (cf. [9], [10]) have the rules MP and Adj as rules of inference but B and Brady's logics have Pref and Suf, and E has the rule Assertion (i.e., $A \Rightarrow (A \to B) \to B$) only as rules of proof.

Rules of inference and rules of proof can be assumed as primitive in a system L or they can be proved as derived rules in L. However, notice that if for example Pref or Suf were added to B as rules of inference, the resulting logic will be considerably stronger.

Rules of proof are present in many of the systems that follow, and, certainly, in all the basic ones. This fact entails that the notion of "proof from premises" cannot be defined as it is customary in Hilbert-style axiomatic systems, to wit, A follows from a set of premises Γ iff there is a finite sequence of wffs, each element of which is a premise, an axiom or the result of applying a rule of derivation to one or more precedent formulas in the sequence, A being the last formula in it: otherwise, the rules of proof in question would in fact be used as rules of inference. Instead, the following is the adequate notion of "proof from premises" in EB_+-logics in which some of its rules are rules of proof.

Definition 1.32 (Proof from premises). Let L be a logic formulated in the Hilbert-style way, one or more rules of proof being included in the set of rules of derivation of L. And let Γ be a set of wffs and A a wff of the language of L. Then, $\Gamma \vdash_L A$ ("A is derivable in L from the set of premises Γ") iff there is a sequence of wffs $B_1, ..., B_n$ such that B_n is A and each B_i $(1 \leq i \leq n)$ is one of the following: (1) $B_i \in \Gamma$; (2) B_i is a theorem of L; (3) B_i is the result of applying the rule Adj to two precedent formulas in the sequence; (4) B_i is the result of applying the rule L-imp (that is, there is some $C \in \mathcal{F}$ such that $C \rightarrow B_i$ is a theorem of L and C is some B_k $(k < i)$).

This notion of "proof from premises" is coextensive ("equivalent") with the following concept of semantical consequence defined in EB_+-models (a corresponding definition for EB_{K+}-models is provided in Definition 3.38).

Definition 1.33 (Semantical consequence relation). Let L, Γ, and A be as in Definition 1.32 and suppose given an RM-semantics for L built upon EB_+-models (cf. Definitions 1.4 and 1.7). Then, $\Gamma \vDash_L A$ ("A is a semantical consequence of Γ in L-semantics") iff for any L-model M and $a \in O$, if $a \vDash \Gamma$, then $a \vDash A$ (for any model M and $a \in X$ $(X \subseteq K)$, $a \vDash \Gamma$ iff $a \vDash B$ for all $B \in \Gamma$).

Then, the following "completeness theorem" (of sorts) can be proved for EB_+-logics with an RM-semantics defined upon EB_+-models.

Theorem 1.34 (Coextensiveness of $\vdash_L$ and $\vDash_L$). *For any set of wffs Γ and wff A, $\Gamma \vdash_L A$ iff $\Gamma \vDash_L A$.*

Let us sketch a proof of this theorem in the case of B_+.

Theorem 1.35 ($\Gamma \vdash_{B_+} A$ iff $\Gamma \vDash_{B_+} A$). *For any set of wffs Γ and wff A, $\Gamma \vdash_{B_+} A$ iff $\Gamma \vDash_{B_+} A$.*

Proof. (a) ($\Rightarrow$) If $A \in \Gamma$ or A is by Adj, the proof is trivial. And if A is a theorem, the proof follows by weak soundness (Corollary 1.29). Finally, suppose that A has been derived by B$_+$-imp. Then, for some $B \in \mathcal{F}$, we have (1) $\vdash_{B_+} B \to A$ and (2) $\Gamma \vdash_{B_+} B$. Let now $a \in O$ in a B$_+$-model and (3) $a \vDash \Gamma$. We have to prove $a \vDash A$. By 2, we have (4) $\Gamma \vDash_{B_+} B$; and by 3 and 4, (5) $a \vDash B$. By d1 and P1, we get (6) $Rxaa$ for some $x \in O$; and by 1 and weak soundness (Corollary 1.29), we derive (7) $\vDash_{B_+} B \to A$, whence by Definition 1.6, (8) $x \vDash B \to A$ follows since $x \in O$. Finally, we have (9) $a \vDash A$, as required, by 5, 6, and 8.

(b) ($\Leftarrow$) For any set of wffs Θ, we define the set of consequences of Θ in B$_+$, CnΘ[B$_+$], as follows: CnΘ[B$_+$] = $\{A \mid \Theta \vdash_{B_+} A\}$. Then, it is trivial to prove that for any Θ, CnΘ[B$_+$] is a regular B$_+$-theory. Now, suppose that Γ is a set of wffs and A a wff such that $\Gamma \nvdash_{B_+} A$. It follows that $A \notin$ CnΓ[B$_+$] whence by applying Lemma 1.17 we obtain a regular prime theory $\mathcal{T}$ such that $A \notin \mathcal{T}$ and CnΓ[B$_+$] $\subseteq \mathcal{T}$. As $\Gamma \subseteq$ CnΓ[B$_+$], we have $\Gamma \subseteq \mathcal{T}$, i.e., $\mathcal{T} \vDash^P \Gamma$ (since $\mathcal{T} \vDash^P B$ for all $B \in \Gamma$). But $\mathcal{T} \nvDash^P A$. That is, $\Gamma \nvDash_{B_+} A$, by applying Definitions 1.27 and 1.33, as it was to be proved. □

Strong soundness and completeness theorems of this form are easily provable (following the pattern set up above for B$_+$) for any of the logics defined in the present work, provided it is endowed with an RM-semantics in the sense of Definition 1.7. However, we have no space here to remark these theorems, let alone to prove them. But, like we said, the reader would not find any difficulty in carrying out this task. (We notice that in some cases, standard strong soundness and completeness theorems are provable. For example, in the case of many of the logics defined in Chapters 6 and 7.)

The chapter is ended with a couple of questions. The first one, related to the presence of rules of proof, concerns the dichotomy reduced/unreduced RM-semantics. Now, when defined with a set of designated points, RM-semantics is essentially classified in two types: reduced and unreduced RM-semantics. The main difference between both types of RM-semantics is that O is a set in the latter, but a singleton in the former. In [37], [9], or [10] it is argued at length why reduced models are preferable when it is possible to define them. Nevertheless, only RM-semantics of the unreduced type are defined in the present work. The reason why this is so is connected with the presence of rules of proof in the basic logics and in many of the systems extending or expanding them. We have no room here to appropriately discuss this matter. Suffice it to say that reduced RM-semantics generally demands that each rule of proof be accompanied by its disjunctive version (for example, the disjunctive version of MP is the rule $C \lor A$ & $C \lor (A \to B) \Rightarrow C \lor B$). But, as it is clear, we cannot in general extend a given logic with the disjunctive version of each rule of proof, since it amounts to considerably strengthen the system thus extended (for instance, this is fairly obvious in the case of B_+).

The second question referred to above is the following. Sometimes we will need to speak not about EB_+-logics or EB_{K+}-logics in general but about extensions or expansions of a particular EB_+-logic or EB_{K+}-logic, L. In this sense, the following definition will be useful.

Definition 1.36 (EL-logics, EL-models). Let L be an EB_+-logic or an EB_{K+}-logic and an L-model be an EB_+-model or an EB_{K+}-model in which postulates Pi_1, ..., Pi_n hold. By an EL-logic we mean an extension or expansion (as the case may be) of L. On the other hand, an EL-model is defined exactly as an L-model except for, possibly, the addition of postulates Pj_1, ..., Pj_m and/or one or more of the clauses (v), (vi), (vii), (viii), (ix), or (x) (cf. Definitions 1.4 and 3.3). The notions of truth and validity in a given class of EL-models are understood according to Definitions 1.5 and 1.6 or Definitions 3.4 and 3.5, as the case may be. In general, EL-validity is validity in any class of EL-models (cf. Definition 1.6).

CHAPTER 2

The basic constructive logics $\mathbf{B}_{cS}$ and $\mathbf{B}_c$

In this chapter, we study a negation expansion of the logic $\mathbf{B}_+$ (cf. Definition 1.10) with the intuitionistic-type negation discussed in the Introduction. We consider two versions of this negation expansion, the logic $\mathbf{B}_{cS}$ and the logic $\mathbf{B}_c$. In $\mathbf{B}_{cS}$ negation is introduced by clause (v); in $\mathbf{B}_c$, by clause (vi) (clauses (v) and (vi) are labelled clause (5¬) and clause (5'¬), respectively, in the Introduction; cf. Definition 0.11). The logic $\mathbf{B}_{cS}$ is the basic constructive logic in RM_1-semantics with a set $S \subseteq K$ (RM-semantics with a set of designated points, O, and a set of consistent points, S). The logic $\mathbf{B}_c$ is the basic constructive logic in RM_1-semantics with $S = K$ (RM-semantics in which all points are consistent and there is a set of designated points). Unless otherwise stated, consistency is understood as w-consistency throughout this book (cf. the paragraph following Definition 0.8). The term constructive has to be understood in the sense "endowed with an intuitionistic-type negation" (cf. §0.5). Most of the chapter is dedicated to the study of $\mathbf{B}_{cS}$. The logic $\mathbf{B}_c$ is easily accessible from $\mathbf{B}_{cS}$.

2.1 THE LOGIC $\mathbf{B}_{cS}$

In this section, the logic $\mathbf{B}_{cS}$ is defined and some of its theorems are proved. The logic $\mathbf{B}_{cS}$ is built upon the positive language of $\mathbf{B}_+$ expanded with the unary connective ¬ (negation).

Definition 2.1 (The logic $\mathbf{B}_{cS}$). The logic $\mathbf{B}_{cS}$ is axiomatized by adding the following axiom and rules to $\mathbf{B}_+$:

$$\text{A7. } \neg A \rightarrow [A \rightarrow \neg(B \rightarrow B)]$$
$$\text{Red. } A \Rightarrow (B \rightarrow \neg A) \rightarrow \neg B$$
$$\text{VeqDn. } A \Rightarrow \neg\neg A \rightarrow (B \rightarrow \neg\neg A)$$

The abbreviation Veq stands for "Verum e quodlibet" —"A true proposition follows from any proposition"—; Dn, for "double negation". Thus,

Routley-Meyer Ternary Relational Semantics for Intuitionistic-Type Negations.
DOI: http://dx.doi.org/10.1016/B978-0-08-100751-8.00003-1

the rule VeqDn is a restriction of the general axiom Veq $A \to (B \to A)$ to the case of doubly negated theorems. On the other hand, the rule Red is one of the forms of the rule reductio (Red stands for 'reductio').

We note that B_{cS} is well-axiomatized w.r.t. B_+. That is, given B_+, A7, Red, and VeqDn are independent from each other (cf. Proposition 2.38). This apparent mere formal question is important and we shall return to it at the end of the chapter.

The following are some theorems and derived rules of B_{cS} (a proof is sketched for each one of them). Part of these theorems and rules are used in the proof of the definitional equivalence between B_{cS} and $B_{cS,f}$ (cf. Chapter 4. In $B_{cS,f}$, negation is introduced by means of the falsity constant f).

T1. $[B \to \neg(A \to A)] \to \neg B$	A1, Red
T2. $\neg B \Rightarrow (A \to B) \to \neg A$	A7, T1
T3. $A \to B \Rightarrow \neg B \to \neg A$	A7, T1
T4. $A \Rightarrow \neg\neg A$	A1, Red
T5. $A \Rightarrow B \to \neg\neg A$	VeqDn, T4
T6. $A \Rightarrow B \to (C \to \neg\neg A)$	VeqDn, T5
T7. $A \Rightarrow \neg(\neg\neg A \to \neg\neg A) \to \neg B$	T3, T6

T8. $A \Rightarrow \neg\neg\neg A \to (B \to \neg C)$

Proof sketch.

1. A	Hyp.
2. $\neg\neg\neg A \to [\neg\neg A \to \neg(\neg\neg A \to \neg\neg A)]$	A7
3. $\neg(\neg\neg A \to \neg\neg A) \to \neg C$	T7, 1
4. $[\neg\neg A \to \neg(\neg\neg A \to \neg\neg A)] \to (\neg\neg A \to C)$	Pref, 3
5. $\neg\neg A \to (\neg\neg A \to \neg\neg C)$	Trans, 2, 4
6. $B \to \neg\neg A$	T5, 1
7. $(\neg\neg A \to \neg C) \to (B \to \neg C)$	Suf, 6
8. $\neg\neg\neg A \to (B \to \neg C)$	Trans, 5, 7

□

T9. $\neg A \to (A \to \neg B)$

Proof sketch.

$$
\begin{array}{lll}
1. & \neg A \to [A \to \neg[\neg\neg(B \to B) \to \neg\neg(B \to B)]] & \text{A7} \\
2. & \neg\neg(B \to B) & \text{A1, T4} \\
3. & \neg[\neg\neg(B \to B) \to \neg\neg(B \to B)] \to \neg B & \text{A1, T7} \\
4. & \neg A \to (A \to \neg B) & 1, 3
\end{array}
$$

$\square$

$$
\begin{array}{lll}
\text{T10.} & A \Rightarrow \neg A \to \neg B & \text{T4, T9} \\
\text{T11.} & A \Rightarrow \neg A \to (B \to \neg C) & \text{T8, T10} \\
\text{T12.} & \neg A \Rightarrow B \to \neg A & \text{A7, T3, T5} \\
\text{T13.} & A \ \& \ B \Rightarrow \neg A \leftrightarrow \neg B & \text{T10} \\
\text{T14.} & A \ \& \ B \Rightarrow (A \to \neg B) \to \neg B & \text{Red, T13}
\end{array}
$$

T15. $A \ \& \ B \Rightarrow (A \to \neg B) \to [C \to (A \to \neg B)]$

Proof sketch.

$$
\begin{array}{lll}
1. & A \ \& \ B & \text{Hyp.} \\
2. & \neg B \to (C \to \neg B) & \text{T11, 1} \\
3. & (A \to \neg B) \to \neg B & \text{T14, 1} \\
4. & (A \to \neg B) \to (C \to \neg B) & \text{Trans, 2, 3} \\
5. & \neg B \to (A \to \neg B) & \text{T11, 1} \\
6. & (C \to \neg B) \to [C \to (A \to \neg B)] & \text{Pref, 5} \\
7. & (A \to \neg B) \to [C \to (A \to \neg B)] & \text{Trans, 4, 6}
\end{array}
$$

$\square$

T16. $A \ \& \ B \Rightarrow [[(A \to \neg B) \to \neg B] \to [C \to [(A \to \neg B) \to \neg B]]]$

Proof sketch.

$$
\begin{array}{lll}
1. & A \ \& \ B & \text{Hyp.} \\
2. & \neg\neg A \to (\neg A \to \neg B) & \text{T9} \\
3. & (A \to \neg B) \leftrightarrow \neg A & \text{T4, Red, 1}
\end{array}
$$

$$4.\ \neg\neg A \to [(A \to \neg B) \to \neg B] \qquad\qquad 2, 3$$
$$5.\ (C \to \neg\neg A) \to [C \to [(A \to \neg B) \to \neg B] \qquad\qquad \text{Pref, } 4$$
$$6.\ [(A \to \neg B) \to \neg B] \to (C \to \neg\neg A) \qquad\qquad \text{T6, } 1$$
$$7.\ [(A \to \neg B) \to \neg B] \to [C \to [(A \to \neg B) \to \neg B] \qquad\qquad \text{Trans, } 5, 6$$

$$\square$$

$$\text{T17. } \neg(A \lor B) \to (\neg A \land \neg B) \qquad\qquad \text{A3, A4, T3}$$
$$\text{T18. } (\neg A \land \neg B) \to \neg(A \lor B) \qquad\qquad \text{A7, Red}$$
$$\text{T19. } (\neg A \lor \neg B) \to \neg(A \land B) \qquad\qquad \text{A2, A5, T3}$$

2.2 AN RM-SEMANTICS FOR $\mathbf{B}_{cS}$

We begin by defining the notion of a B_{cS}-model and related notions.

Definition 2.2 (B_{cS}-models). A B_{cS}-model is an EB$_+$-model $(K, O, S, R, \vDash)$ where clauses (i)–(v) and the following semantical postulates (in addition to P1 and P2) are satisfied for all $a, b, c \in K$:

$$\text{P3. } (Rabc\ \&\ c \in S) \Rightarrow a \in S$$
$$\text{P4. } (Rabc\ \&\ c \in S) \Rightarrow \exists x \in O\ \exists y \in S\ Rbxy$$

The notions of truth in a B_{cS}-model and Bcs-validity are understood according to the general Definitions 1.5 and 1.6 (cf. also Definition 1.4).

We note the following proposition:

Proposition 2.3 (Postulates holding in B_{cS}-models). *The following semantical postulates hold in all B_{cS}-models:*

$$\textit{Pa. } a \in S \Rightarrow \exists x \in O\ \exists y \in S\ Raxy$$
$$\textit{Pb. } (Rabc\ \&\ c \in S) \Rightarrow b \in S$$
$$\textit{Pc. } (a \leq b\ \&\ b \in S) \Rightarrow a \in S$$
$$\textit{Pd. } (Rabc\ \&\ c \in S) \Rightarrow \exists x \in O\ \exists y \in S\ Rcxy$$

Proof. Pa: by P1, P4, and d1; Pb: by P3 and P4; Pc: by Pb and d1; Pd: by Pa. $\qquad\square$

Extensions or expansions of B_{cS}, EB_{cS}-logics, as well as the notions of an EB_{cS}-model and EB_{cS}-validity are understood according to Definition 1.36.

As it was the case with B_+, we shall proceed to the proof of the soundness and completeness of B_{cS} in a general way in order to use the results obtained in the soundness and completeness proofs of logics including B_{cS}.

Proposition 2.4 (All theorems of B_{cS} are EB_{cS}-valid)**.** *For each formula A, if $\vdash_{B_{cS}} A$, then A is EB_{cS}-valid.*

Proof. (We use the Entailment Lemma —Lemma 1.9.) Given that Proposition 1.13 guarantees that all axioms of B_+ (A1–A6) are EB_{cS}-valid and the rules of B_+ (MP, Adj, Suf, and Pref) preserve EB_{cS}-validity, it remains to be proved that A7 is EB_{cS}-valid and that the rules Red and VeqDn preserve EB_{cS}-validity (recall that EB_{cS}-validity means validity in any class of EB_{cS}-models and that by i, ii, etc., we refer to clauses (i)–(v) in Definition 1.4).

(a) *A7, $\neg A \to [A \to \neg(B \to B)]$, is EB_{cS}-valid:* Suppose that A7 is not EB_{cS}-valid. By Lemma 1.9, there are $A, B \in \mathcal{F}$ and $a \in K$ in some EB_{cS}-model M such that (1) $a \vDash \neg A$ and (2) $a \nvDash A \to \neg(B \to B)$. Then, by 2 and iv, there are $b, c \in K$ such that (3) $Rabc$, (4) $b \vDash A$, and (5) $c \nvDash \neg(B \to B)$. By 5 and v, there are $d, e \in K$ in M such that (6) $Rcde$, (7) $e \in S$ and (8) $d \vDash B \to B$. By P3, 6, and 7, we get (9) $c \in S$; and by 1, 3, 9, and v, we have (10) $b \nvDash A$, contradicting 4. Therefore, A7 is EB_{cS}-valid.

(b) *Red, $A \Rightarrow (B \to \neg A) \to \neg B$, preserves EB_{cS}-validity:* Suppose that there are $A, B \in \mathcal{F}$ such that (1) $\vDash A$ but (2) $\nvDash (B \to \neg A) \to \neg B$. By Lemma 1.9 and 2, there is $a \in K$ in some EB_{cS} model M such that (3) $a \vDash B \to \neg A$ and (4) $a \nvDash \neg B$. By 4 and v, there are $b \in K$ and (5) $c \in S$ in M such that (6) $Rabc$ and (7) $b \vDash A$. By 3, 6, 7, and iv, we get (8) $c \vDash \neg A$. Now, given 5, 6, and Pd (cf. Proposition 2.3) (9) there are $x \in O$ and $y \in S$ such that $Rcxy$. Thus, we have (10) $x \nvDash A$ by 8 and 9. But, on the other hand, we have (11) $x \vDash A$ by 1, since $x \in O$ by 9. Consequently, Red preserves EB_{cS}-validity.

(c) *VeqDn, $A \Rightarrow \neg\neg A \to (B \to \neg\neg A)$, preserves EB_{cS}-validity:* Supposing that there are $A, B \in \mathcal{F}$ such that (1) $\vDash A$ but (2) $\nvDash \neg\neg A \to (B \to \neg\neg A)$, we have, by Lemma 1.9 and 2, (3) $a \vDash \neg\neg A$ and (4) $a \nvDash B \to \neg\neg A$, for some $a \in K$ in some EB_{cS}-model M, whence by iv, we get (5) $Rabc$, (6) $b \vDash B$ and (7) $c \nvDash \neg\neg A$ for some $b, c \in K$ in M. By 7 and v, we have (8) $Rcde$ and (9) $d \vDash \neg A$ for some $d \in K$ and $e \in S$. Now, by P4, 8, and 9, we get (10) there are $x \in O$ and $y \in S$ such that $Rdxy$. So, (11) $x \nvDash A$ follows by 9,

10, and v. But, on the other hand, we have (12) $x \vDash A$ by 1 since $x \in O$. Therefore, VeqDn preserves EB_{cS}-validity.   □

Corollary 2.5 (Soundness of B_{cS}). *For any formula A, if $\vdash_{B_{cS}} A$, then $\vDash_{B_{cS}} A$.*

Proof. Immediate by Proposition 2.4 since a B_{cS}-model is an EB_{cS}-model.   □

2.3 COMPLETENESS OF B_{cS} I. ON W-CONSISTENCY

In the rest of this chapter, by L we generally refer to an EB_{cS}-logic (not simply to an EB_+-logic), unless otherwise stated (cf. Definitions 1.14, 1.15, and 1.36).

In this section, we prove some facts about w-consistency in the context of EB_{cS}-logics, some of which will be used in the completeness proofs of the sections to follow (we recall that an L-theory a is w-inconsistent iff it contains the negation of a theorem of L and that a is w-consistent iff it is not w-inconsistent —cf. Definition 1.15).

Before going into this analysis of w-consistency, we note the following remark and a property of non-empty theories.

Remark 2.6 (On the w-consistency of EB_+-logics). Let $\mathcal{L}$ be a language with $\rightarrow$, $\wedge$, $\vee$, and $\neg$ as the sole propositional connectives. All logics built upon $\mathcal{L}$ treated in the present work are axiomatized by axioms and rules that are provable in classical propositional logic when the connectives $\rightarrow$, $\wedge$, $\vee$, and $\neg$ are read as the classical propositional connectives. Consequently, all logics built upon $\mathcal{L}$ studied in the present book are w-consistent.

Proposition 2.7 (On non-empty theories). *Let $a \in K^{NT}$ and $A, \neg B$ be theorems of L. Then, $\neg\neg A \in a$ and $\neg B \in a$.*

Proof. Immediate by T5 and T12, respectively.   □

Next, we prove the aforementioned facts about w-consistency.

Proposition 2.8 (On w-consistency and negative formulas). *Let $a \in K^T$. Then, a is w-inconsistent iff a contains all negative formulas.*

Proof. (a) ($\Rightarrow$) Suppose $\neg A \in a$, A being a theorem. By T10, $\neg A \rightarrow \neg B$ is a theorem for any $B \in \mathcal{F}$. So, we have $\neg B \in a$, for arbitrary B. (b) ($\Leftarrow$) It is obvious.   □

A corollary of this proposition is the following:

Proposition 2.9 (W-consistency and negation of theorems). *Let $a \in K^T$. Then, a is w-inconsistent iff a contains the negation of every theorem of L.*

Proof. Immediate by Proposition 2.8.   □

   Two more facts are worth recording. The first one states the equivalence of w-consistency and u-consistency in the context of EB_{cS}-logics. We recall that an L-theory a is u-inconsistent iff $A \in a$, $\neg A$ being a theorem of L (a is u-consistent iff it is not u-inconsistent).

Proposition 2.10 (W-consistency and u-consistency). *Let $a \in K^T$. Then a is w-consistent iff a is u-consistent.*

Proof. (a) ($\Rightarrow$) Suppose a is w-inconsistent, that is, $\neg A \in a$, A being a theorem. By T4, $\neg\neg A$ is a theorem. So, a is u-inconsistent. (b) ($\Leftarrow$) Suppose $A \in a$, $\neg A$ being a theorem. By T9, we have $\vdash_{B_{cS}} A \to \neg C$, C being a theorem. So, $\neg C \in a$ follows and hence a is w-inconsistent.   □

Proposition 2.11 (W-consistency and contradictions). *Let $a \in K^T$. If a is w-inconsistent, then a contains a contradiction.*

Proof. Let $\neg A \in a$, A being a theorem. By Proposition 2.7, we have $\neg\neg A \in a$. So, we get $\neg A \wedge \neg\neg A \in a$.   □

   Concerning Proposition 2.11, we note the following remark.

Remark 2.12 (On Proposition 2.11). The converse of Proposition 2.11 is neither provable in B_{cS} nor in many of its extensions. On the other hand, notice that any trivial L-theory is w-inconsistent but that not every w-inconsistent L-theory is necessarily trivial (cf. §7.3 in Chapter 7).

   The section is ended with a proof of a second primeness lemma. So far, we have a lemma (Lemma 1.17) showing how to extend non-trivial L-theories to prime ones. Next, it is proved how w-consistent L-theories can be extended to prime w-consistent L-theories.

Lemma 2.13 (Extension to prime, w-consistent L-theories). *Let $a \in S^T$. Then, there is $x \in S^P$ such that $a \subseteq x$.*

Proof. We build from a a maximal w-consistent L-theory x such that $a \subseteq x$. Suppose x is not prime. Then, there are $A, B \in \mathcal{F}$ such that $A \vee B \in x$, $A \notin x$, and $B \notin x$. Similarly as in Lemma 1.17, we define the sets $[x, A]$ and $[x, B]$ and prove that they are L-theories strictly including x. By the maximality of x, they are w-inconsistent, that is, $\neg C \in [x, A]$, $\neg D \in [x, B]$ for some theorems C and D of L. Then, we have $\vdash_L (A \wedge E) \to \neg C$, $\vdash_L (B \wedge E') \to \neg D$, for some $E \in x$, $E' \in x$, whence by using E$\vee$2, t2, and trans of FDE$_+$ (cf. Proposition 1.11), we get $\vdash_L [(A \vee B) \wedge (E \wedge E')] \to (\neg C \vee \neg D)$, and so, $\neg C \vee \neg D \in x$ (as $(A \vee B) \wedge (E \wedge E') \in x$). Finally, we have $\neg(C \wedge D) \in x$ by applying T19. But by Adj, $C \wedge D$ is a theorem of L. Then x is w-inconsistent, which is impossible. Therefore, x is prime.   □

Remark 2.14 (On Lemma 2.13). We remark that Lemma 2.13 is provable in any EB$_+$-logic in which the rule Contraposition (Con: $A \to B \Rightarrow \neg B \to \neg A$) or the rule De Morgan (DM: $\neg A \vee \neg B \Rightarrow \neg(A \wedge B)$) hold. (The rule DM is immediate by A2, A5, and Con.) In order to show this fact, T13 (if A and B are theorems, then $\neg A$ and $\neg B$ are equivalent) has not been used in the proof of Lemma 2.13: given $\vdash_L (A \wedge E) \to \neg C$, $\vdash_L (B \wedge E') \to \neg D$ and T13, we have, $\vdash_L (B \wedge E') \to \neg C$. Then, we get $\vdash_L [(A \vee B) \wedge (E \wedge E')] \to \neg C$, whence $\neg C \in x$, and x is w-inconsistent.

2.4 COMPLETENESS OF B$_{cS}$ II. THE CANONICAL MODEL. THE COMPLETENESS THEOREM

In this section, we will prove the completeness of B$_{cS}$ w.r.t. the semantics defined in §2.2. We start with a series of lemmas that will prove useful not only here, but also in other sections to follow. (Maybe, at this point, it is convenient to consult Definition 1.16.)

Lemma 2.15 (R^T and non-emptiness). *Let $a, b \in K^{NT}$ and $c \in K^T$ such that $R^T abc$. Then, $c \in K^{NT}$.*

Proof. Let $A \in a$, $B \in b$ and C be a theorem of L. By T6, $A \to (B \to \neg\neg C)$ is a theorem of B$_{cS}$. So, we have $B \to \neg\neg C \in a$ and, finally, $\neg\neg C \in c$. Thus, c is non-empty.   □

Remark 2.16 (On Lemma 2.15). Notice that a theorem with the general structure $A \to (B \to C)$ $(A \neq B)$ suffices for proving the simple but important Lemma 2.15. Here we use T6, which is not provable in B$_{cS}$ if the rule VeqDn is dropped. We shall return to this question in §2.6.

Lemma 2.17 (The pre-canonical P3). *Let $a, b \in K^{NT}$ and $c \in S^T$ such that $R^T abc$. Then, $a, b \in S^{NT}$.*

Proof. (a) Suppose $\neg A \in a$, A being a theorem of L, and let $B \in b$ and C be a theorem. By T11, $\neg A \to (B \to \neg C)$ is a theorem. Then, we have $B \to \neg C \in a$, and so, $\neg C \in c$, contradicting the w-consistency of c. Therefore, a is w-consistent. (b) Suppose $\neg A \in b$, A being a theorem of L. By A7, $\neg\neg A \to [\neg A \to \neg(A \to A)]$ is a theorem of L, and by Proposition 2.7, we get $\neg\neg A \in a$. So, $\neg A \to \neg(A \to A) \in a$ follows, whence $\neg(A \to A) \in c$, contradicting the w-consistency of c. Thus, b is w-consistent. $\square$

Lemma 2.18 (The pre-canonical P4). *Let $a, b \in K^{NT}$ and $c \in S^T$ such that $R^T abc$. Then, there are $x, z \in O^P$ and $y, u \in S^{NP}$ such that (a) $R^{NP} bxy$ and (b) $R^{NP} czu$.*

Proof. We prove case (a) (the proof of (b) is similar). Given the hypothesis of Lemma 2.18, we define the sets $z = \{A \mid \vdash_L A\}$, $u = \{B \mid \exists A[A \to B \in b \ \& \ A \in z]\}$. It is obvious that z is an L-theory and it is easy to prove that u is an L-theory such that $R^T bzu$ (cf. the proof of Lemma 1.25e). Clearly, $z \in O^T$ while $u \in K^{NT}$ (by Lemma 2.15). It remains to prove that u is w-consistent. Suppose it is not. Then, we have $\neg A \in u$, A being a theorem. So, $B \to \neg A \in b$ follows, B being a theorem (by definition of z and u). Thus, we get $\neg B \in b$ by Red. On the other hand, $\neg\neg B \to [\neg B \to \neg(B \to B)]$ is a theorem by A7, and we have $\neg\neg B \in a$ by Proposition 2.7. Therefore, we obtain $\neg B \to \neg(B \to B) \in a$ and, finally, $\neg(B \to B) \in c$ (as $R^T abc$), contradicting the w-consistency of c. Consequently, we have $z \in O^T$ and $u \in S^{NT}$ such that $R^{NT} bzu$. Next, z and u are extended to the required prime L-theories. By Lemma 2.13, u is extended to some $y \in S^{NP}$. Obviously, $R^{NT} bzy$. Then, by Lemma 1.19, z is extended to some $x \in O^P$ such that $R^{NP} bxy$. Thus, we have $x \in O^P$ and $y \in S^{NP}$ such that $R^{NP} bxy$, as it was required. $\square$

Remark 2.19 (On R^T and w-consistency). Let a, b be non-empty and w-consistent L-theories and c an L-theory such that $R^T abc$. Notice that although c is non-empty (Lemma 2.15), it is not necessarily w-consistent.

The ensuing corollary adapts Lemma 1.25e to the needs of EB_{cS}-logics.

Corollary 2.20 (The pre-canonical $\vDash^P$. Conditionals in K^{NP}). *Let $A, B \in \mathcal{F}$ and $a \in K^{NP}$ such that $A \to B \notin a$. Then, there are $x, y \in K^{NP}$ such that $R^{NP} axy$, $A \in x$ and $B \notin y$.*

Proof. Assume the hypothesis of Corollary 2.20. By Lemma 1.25e, there are $x, y \in K^P$ such that $R^P axy$, $A \in x$, and $B \notin y$. It remains to prove that y is non-empty, which is immediately secured by Lemma 2.15 given that a and x are non-empty elements in K^P. $\square$

As regards the following lemmas, Lemma 2.21 is proved in any expansion of B_+ in which A7 in the restricted form A7' $\neg A \to [A \to \neg(A \to A)]$ is a theorem, and Lemma 2.22 holds in any expansion of B_+ in which the rule Red is provable. These lemmas extend Lemma 1.25 with the case of negative formulas.

Lemma 2.21 (The pre-canonical $\vDash^P$. Negation I). *Let L be an EB_+-logic in which A7' $\neg A \to [A \to \neg(A \to A)]$ is provable. For any $A \in \mathcal{F}$ and $a \in K^T$, if $\neg A \in a$, then for any $b, c \in K^T$, $(R^T abc \ \& \ c \in S^T) \Rightarrow A \notin b$.*

Proof. Suppose $\neg A \in a$ and $R^T abc$ for $a, b \in K^T$ and $c \in S^T$. Moreover, suppose, for reductio, $A \in b$. By A7', we have $A \to \neg(A \to A) \in a$ and, finally, $\neg(A \to A) \in c$, contradicting the w-consistency of c. $\qquad\square$

Lemma 2.22 (The pre-canonical $\vDash^P$. Negation II). *Let L be an EB_+-logic in which Red is provable. For any $A \in \mathcal{F}$ and $a \in K^T$, if $\neg A \notin a$, then there are $x \in K^{NP}$ and $y \in S^P$ such that $R^P axy$ and $A \in x$.*

Proof. Suppose $\neg A \notin a$ for $A \in \mathcal{F}$ and $a \in K^T$. Consider the sets $z = \{B \mid\vdash_L A \to B\}$ and $u = \{C \mid \exists D[D \to C \in a \ \& \ D \in z]\}$. They are L-theories such that $R^T azu$ (cf. the proof of case e in Lemma 1.25) and $A \in z$ (by A1, $A \to A$ is a theorem of L). It remains to prove $u \in S^T$. Suppose that u is not w-consistent. That is, suppose $\neg B \in y$, B being a theorem of L. Then, we have $C \to \neg B \in a$, $\vdash_L A \to C$ for some $C \in \mathcal{F}$. By the rule Suf, we get $\vdash_L (C \to \neg B) \to (A \to \neg B)$. So, $A \to \neg B \in a$ follows, whence, by Red, we get $\neg A \in a$, contradicting the hypothesis. Consequently, we have $x \in K^{NT}$ and $u \in S^T$ such that $R^T azu$ and $A \in z$. Next, z and u are extended to the required prime L-theories x and y similarly as in Lemma 2.18. $\qquad\square$

Consider the following definition.

Definition 2.23 ($a \leq^S b$). Let $a \in K^P$ and $b \in S^P$. $a \leq^S b$ iff $\exists x \in O^P \cap S^P$ $R^P xab$.

The lemma that follows will be useful in some EB_{cS}-logics.

Lemma 2.24 ($a \leq^S b$ iff $a \subseteq b$). *For any $a \in K^P$, $b \in S^P$, $a \leq^S b$ iff $a \subseteq b$.*

Proof. Let $a \in K^P$, $b \in S^P$. (a) ($\Rightarrow$) Suppose $a \leq^S b$. By Corollary 1.23, we get $a \subseteq b$. (b) ($\Leftarrow$) Suppose $a \subseteq b$. By Corollary 1.23 and Definition 1.21, there is $x \in O^P$ such that $R^P xab$. By Lemma 2.17 x is w-consistent, i.e., $x \in S^P$. Then, we have $a \leq^S b$ by Definition 2.23. $\qquad\square$

In what follows, we proceed to the proof of the completeness of B_{cS}. Firstly, the canonical model is defined.

Definition 2.25 (The canonical B_{cS}-model). The canonical B_{cS}-model is a canonical EB_+-model of type c (cf. Definition 1.26). That is, it is the structure $(K^{NP}, O^P, S^{NP}, R^{NP}, \vDash^{NP})$. (Recall that K^{NP}, O^P, S^{NP} and $\vDash^{NP}$ are here referred to B_{cS}-theories.)

Remark 2.26 (On the canonical B_{cS}-model). The canonical B_{cS}-model (as well as the canonical models for EB_{cS}-logics in general) are clearly distinguishable from canonical models for standard relevance logics or from those for such logics as intuitionistic logics or Lewis' modal logics. Similarly as in relevance logics, not all theories have to be consistent in any sense of the term. But unlike in relevance logics, all theories have to be non-empty. Finally, unlike in minimal intuitionistic logic or Lewis' S4 (where non-empty and regular are coextensive terms), not all theories need to be regular.

We will prove the completeness of B_{cS} following the general strategy sketched in Remark 1.31.

Lemma 2.27 (The canonical model is indeed a model). *The canonical B_{cS}-model is in fact a B_{cS}-model.*

Proof. We prove the facts (1)–(3) listed below. (Cf. Lemma 1.28.)
1. The set $O^P \cap S^{NP}$ is not empty. It is immediate by Lemma 2.13, since B_{cS} is, of course, a regular and w-consistent B_{cS}-theory.
2. Postulates P1–P4 hold in the canonical B_{cS}-model. As it was the case in B_+, P1 and P2 are immediate by Corollary 1.23 (notice that this lemma suffices here, in the context of the canonical B_{cS}-model). Then, P3 and P4 are proved as follows.
    (a) P3, $(R^{NP}abc \ \& \ c \in S^{NP}) \Rightarrow a \in S^{NP}$: Suppose $R^{NP}abc$, for $a, b \in K^{NP}$ and $c \in S^{NP}$. By Lemma 2.17, a is w-consistent. So $a \in S^{NP}$.
    (b) P4, $(R^{NP}abc \ \& \ c \in S^{NP}) \Rightarrow \exists x \in O^P \ \exists y \in S^{NP} \ R^{NP}bxy$: Immediate by Lemma 2.18(a).
3. Clauses (i)–(iv) and (v) in Definition 2.2 are satisfied by the canonical B_{cS}-model. Clauses (i)–(iii) and (iv) (from left to right) are proved similarly as in Lemma 1.28. Then, clause (iv) (from right to left) is immediate by Corollary 2.20. On the other hand, clause (v) (for $A \in \mathcal{F}$ and $a \in K^{NP}$, $\neg A \in a$ iff for $b, c \in K^{NP}$, $(R^{NP}abc \ \& \ c \in S^{NP}) \Rightarrow A \notin b$ —cf. Definitions 2.2 and 2.25) is proved as follows:
    (a) ($\Rightarrow$) Suppose, for $A \in \mathcal{F}$ and $a, b, c \in K^{NP}$, $\neg A \in a$, $R^{NP}abc$, and $c \in S^{NP}$. Then, $A \notin b$ is immediately derivable by Lemma 2.21.

(b) ($\Leftarrow$) Suppose there is $A \in \mathcal{F}$ and $a \in K^{NP}$ such that $\neg A \notin a$. By Lemma 2.22, there are $x \in K^{NP}$ and $y \in S^P$ such that $R^P a x y$ and $A \in x$. By Lemma 2.15, $y \in S^{NP}$. Thus, we have $x \in K^{NP}$ and $y \in S^{NP}$ such that $R^{NP} a x y$, as required.

$\square$

Finally, we prove the (weak) completeness theorem (cf. Remark 1.31).

Theorem 2.28 (Completeness of B$_{cS}$)**.** *For any $A \in \mathcal{F}$, if $\vDash_{B_{cS}} A$, then $\vdash_{B_{cS}} A$.*

Proof. Suppose $\nvdash_{B_{cS}} A$. By Lemma 1.17, there is a regular, prime B$_{cS}$-theory x such that Bcs $\subseteq x$ and $A \notin x$. By Definition 2.25 and Lemma 2.27, $x \nvDash^C A$. Therefore, $\nvDash_{B_{cS}} A$ by Definition 2.2. $\square$

2.5 THE LOGIC B$_c$

As the logic B$_{cS}$, the logic B$_c$ (basic constructive logic in RM-semantics with a set of designated points but not selected set S of consistent points —cf. §2.1) is built upon the positive language of B$_+$ expanded with the unary connective $\neg$ (negation). B$_c$ is defined as follows.

Definition 2.29 (The logic B$_c$)**.** The logic B$_c$ is axiomatized by adding the following axiom and rules to B$_+$:

$$A8. \ \neg A \to (A \to B)$$
$$\text{Red.} \ A \Rightarrow (B \to \neg A) \to \neg B$$
$$\text{VeqDn.} \ A \Rightarrow \neg\neg A \to (B \to \neg\neg A)$$

Thus, B$_c$ is formulated by strengthening A7 to the unrestricted A8, Heyting's tenth axiom for propositional intuitionistic logic (cf., the heads "Intuitionistic Logic" and related topics in [16] and [44]). Consequently, B$_{cS}$ is a sublogic of Bc and T1–T19 of B$_{cS}$ are also provable in B$_c$ (we shall return to the relationship between B$_{cS}$ and B$_c$ at the end of this chapter).

Next, we turn to the RM-semantics for B$_c$.

Definition 2.30 (B$_c$-models)**.** A B$_c$-model is an EB$_+$-model $(K, O, R, \vDash)$ where clauses (i)–(iv) and (vi) and the following semantical postulate (in addition to P1 and P2) are satisfied:

$$P5. \ Rabc \Rightarrow \exists x \in O \ \exists y \in K \ Rbxy$$

The notions of truth in a B_c-model and B_c-validity are understood according to the general Definitions 1.5 and 1.6 (cf. also Definition 1.4).

The ensuing additional postulates can be useful (cf. Proposition 2.3).

Proposition 2.31 (Postulates holding in B_c-models). *The following semantical postulates hold in all B_c-models:*

$$Pe.\ a \in K \Rightarrow \exists x \in O\ \exists y \in K\ Raxy$$
$$Pg.\ Rabc \Rightarrow \exists x \in O\ \exists y \in K\ Rcxy$$

Proof. Pe: by P1, P5, and d1; Pg: by Pe. □

In what follows, we proceed to the proof of the soundness and completeness of B_c according to the general pattern set above for B_+ and B_{cS} (cf. Remark 1.31). (The notion of an EB_c-model and related notions are understood according to the general Definition 1.36.)

Proposition 2.32 (All theorems of B_c are EB_c-valid). *For each $A \in \mathcal{F}$, if $\vdash_{B_c} A$, then A is EB_c-valid.*

Proof. The axioms and rules of B_+ are taken care of as in B_{cS} (Proposition 2.4). The rules Red and VeqDn are proved similarly as in B_{cS} (Proposition 2.4) omitting all references to the set S. Let us show it by proving, e.g., Red.

(a) The rule Red, $A \Rightarrow (B \to \neg A) \to \neg B$, preserves EB_c-validity: Suppose that there are $A, B \in \mathcal{F}$ such that (1) $\vDash A$ but (2) $\nvDash (B \to \neg A) \to \neg B$. By Lemma 1.9 and 2, we have, for $a \in K$ in some EB_c-model M, (3) $a \vDash B \to \neg A$ and (4) $a \nvDash \neg B$. By 4 and vi, there are $b, c \in K$ such that (5) $Rabc$ and (6) $b \vDash A$. By 3, 5, and 6, we get (7) $c \vDash \neg A$. Now, given 5 and Pg (cf. Proposition 2.31), (8) $Rcxy$ follows for $x \in O$ and $y \in K$. Thus, we have (9) $x \nvDash A$ by 7 and 8. But, on the other hand, we have (10) $x \vDash A$ by 1, since $x \in O$ by 8. Consequently, Red preserves EB_c-validity.

Finally, we prove:

(b) A8, $\neg A \to (A \to B)$, is EB_c-valid: Suppose there are $A, B \in \mathcal{F}$ and $a \in K$ in some EB_c-model M such that (1) $a \vDash \neg A$ but (2) $a \nvDash A \to B$. By 2 and iv, there are $b, c \in K$ in M such that (3) $Rabc$, (4) $b \vDash A$, and (5) $c \nvDash B$. But by 1, 3, and vi, we have (6) $b \nvDash A$, contradicting 4. □

Corollary 2.33 (Soundness of Bc). *For $A \in \mathcal{F}$, if $\vdash_{B_c} A$, then $\vDash_{B_c} A$.*

Proof. Immediate by Proposition 2.32. □

In the following two pages, we turn to the completeness proof of B_c. (Recall that, unless otherwise stated, by L we generally mean an EB_{cS}-logic, not simply an EB_+-logic —cf. the opening paragraph of §2.3.)

The canonical model is defined as follows.

Definition 2.34 (The canonical B_c-model). The canonical B_c-model is a canonical EB_+-model of type d (cf. Definition 1.26). That is, it is the structure $(K^{WNP}, O^{WP}, R^{WNP}, \vDash^{WNP})$. (Recall that K^{WNP}, O^{WP}, R^{WNP}, and $\vDash^{WNP}$ are here referred to B_c-theories.)

Given Definitions 2.25 and 2.34, it is clear that the main difference between the canonical B_{cS}-model and the canonical B_c-model is that all theories are w-consistent in the latter while only a subset of all theories need to be w-consistent in the former. The requisite that all theories be w-consistent in the canonical B_c-model has an immediate consequence in the development of the completeness proof for B_c: each new theory introduced has to be shown w-consistent. Nevertheless, the lemmas and propositions proved in the preceding section have been so designed so as to be sufficient not only for EB_{cS}-logics, but also for EB_c-logics. Actually, the only modification required in the said propositions and lemmas in order to establish the completeness of B_c is the following strengthening of Lemma 1.25e (notice that A8 is required: T9 is insufficient).

Lemma 2.35 (The pre-canonical $\vDash^P$. Conditionals in K^{WNP}). *For $A, B \in \mathcal{F}$ and $a \in K^{WNP}$, if $A \rightarrow B \notin a$, then there are $x, y \in K^{WNP}$ such that $R^{WNP}axy$, $A \in x$ and $B \notin y$.*

Proof. Suppose that there are $A, B \in \mathcal{F}$ and $a \in K^{WNP}$ such that $A \rightarrow B \notin a$. By Corollary 2.20, there are $x, y \in K^{NP}$ such that $R^{NP}axy$, $A \in x$, and $B \notin y$. Next, it is shown that y is w-consistent. Suppose it is not. Then, $\neg C \in y$ follows, C being a theorem of B_c. By A8, we have $\vdash_{B_c} \neg\neg C \rightarrow (\neg C \rightarrow B)$, and by T4, $\vdash_{B_c} \neg\neg C$. So we have $\vdash_{B_c} \neg C \rightarrow B$, whence $B \in y$, a contradiction. Consequently, y is w-consistent. Next, x is w-consistent by Lemma 2.17. Thus, we have $x, y \in K^{WNP}$ such that $R^{WNP}axy$, $A \in x$ and $B \notin y$, as it was required. □

Once proved that the canonical model is a model, we can prove completeness.

Lemma 2.36 (The canonical model is a model). *The canonical B_c-model is in fact a B_c-model.*

Proof. (Cf. the proofs of Lemma 1.28 and Lemma 2.27.) We prove the facts (1)–(3) listed below.

1. The set O^{WP} is not empty: By Lemma 2.13, given that Bc is a regular w-consistent theory.

2. Postulates P1–P4 hold in the canonical Bc-model. P1 and P2 are immediate by Lemma 2.24, the required modification of Corollary 1.23 (recall that this corollary was sufficient in the case of B_{cS}, but not here where all theories need to be w-consistent). Then, P5 is proved by Lemma 2.18, similarly as P4 in the canonical B_{cS}-model: Let us see it. Suppose $R^{WNP}abc$ for $a, b, c \in K^{WNP}$. By Lemma 2.18(a), there are $x \in O^P$ and $y \in S^{NP}$ such that $R^{NP}bxy$. Then, x is w-consistent by Lemma 2.17, whence $R^{NP}bxy$ is actually $R^{WNP}bxy$.

3. Clauses (i)–(iv) and (vi) in Definition 2.30 are satisfied by the canonical Bc-model. Clauses (i), (ii), (iii), and (iv) (from left to right) are proved similarly as in Lemma 1.28 (cf. also Lemma 2.27). Then, clause (iv) from right to left follows by Lemma 2.35. Finally, clause (vi) follows from Lemma 2.21 and Lemma 2.22 in a similar way to which clause (v) is provable in the canonical B_{cS}-model (cf. Lemma 2.27).   □

Theorem 2.37 (Completeness of B_c). *For any $A \in \mathcal{F}$, if $\vDash_{B_c} A$, then $\vdash_{B_c} A$.*

Proof. By Lemmas 2.13 and 2.36, and Definitions 2.30 and 2.34 (cf. Remark 1.31).   □

We end the section by recalling that strong soundness and completeness theorems (of sorts) can be proved for B_{cS} and B_c (cf. §1.5).

2.6 B_{cS} AND B_c AS THE BASIC CONSTRUCTIVE LOGICS IN RM_1-SEMANTICS

To end this chapter, we discuss in what sense B_{cS} (respectively, B_c) is the basic constructive logic in RM_1-semantics with a set $S \subseteq K$ (respectively, $S = K$). First of all, we note a couple of propositions.

Proposition 2.38 (Independence in B_{cS} and B_c). *The logic B_{cS} (respectively, B_c) is well-axiomatized w.r.t. B_+. That is, given B_+, A7, Red, and VeqDn (respectively, A8, Red, and VeqDn) are independent from each other.*

Proof. Consider the following sets of truth-tables where designated values are starred. The three sets verify all axioms and rules of B_+. In addition,

Set I verifies A8 and Red; Set II, Red and VeqDn, and Set III, A8 and VeqDn.

Set I (Independence of VeqDn):

→	0	1	2	¬		∧	0	1	2		∨	0	1	2
0	1	1	1	1		0	0	0	0		0	0	1	2
*1	0	1	1	0		*1	0	1	1		*1	1	1	2
*2	0	0	1	0		*2	0	1	2		*2	2	2	2

Falsifies VeqDn when $A = 1, B = 2$.

Set II (Independence of A7):

The tables for ∧ and ∨ are the same as in Set I. The tables for → and ¬ are as follows

→	0	1	2	¬
0	1	1	2	2
*1	0	1	2	0
*2	0	0	2	0

Falsifies A7', $\neg A \rightarrow [A \rightarrow \neg(A \rightarrow A)]$, when $A = 0$.

Set III (Independence of Red):

→	0	1	¬		∧	0	1		∨	0	1
0	1	1	0		0	0	0		0	0	1
*1	0	1	0		*1	0	1		*1	1	1

Falsifies Red when $A = 1$ and $B = 0$.   □

Proposition 2.39 (Unprovability of $A \rightarrow (B \rightarrow C)$). *Let A, B be formulas such that $A \neq B$. Then, there is not a wff C such that $A \rightarrow (B \rightarrow C)$ is provable in B_+ expanded with A8 and Red.*

Proof. We use Set I in Proposition 2.38, which verifies A8 and Red. It suffices to assign the value 2 to both A and B. Then, the value of $A \rightarrow (B \rightarrow C)$ is 0 no matter the value of C. (Notice that Set I also falsifies T6 when $A = B = 2$ and $C = 1$ or 2 —cf. Remark 2.16.)   □

Once these propositions are proved, that B_{cS} is the basic constructive logic in RM_1-semantics with a set $S \subseteq K$ can be established as follows. (1) As shown in Lemmas 2.21 and 2.22, A7' and the rule Red are needed to prove the canonical validity of clause (v) (cf. Definition 1.26 on the canonical definition of the valuation relation ⊨). (2) The validity of A7' demands the semantical postulate P3 (although P3 validates T9, a strong general-ization of A7 —cf. Proposition 2.4). (3) The canonical validity of P3 re-quires non-emptiness of theories in the canonical model (cf. Lemma 2.17). (4) Non-emptiness of theories needs, in its turn, Lemma 2.15. And, as pointed out above, this lemma requires the presence in the logic of a theo-rem of the form $A \rightarrow (B \rightarrow C)$ $(A \neq B)$ where the metavariables A and B represent general formulas (of no particular structure) by means of which one can refer to unspecified formulas occurring in any theories. But, as shown in Proposition 2.39, a theorem of this form is not provable in B_+ plus A8 and Red. (5) It must then be concluded that some particular structure of the form $A \rightarrow (B \rightarrow C)$ (with the required conditions) has to be added as an extra axiom to B_+ expanded with A7 and Red (respectively, A8 and Red). The first schemes coming to mind are (cf. Chapter 6): t27, $A \rightarrow (B \rightarrow A)$; t25, $B \rightarrow (A \rightarrow A)$; t30, $A \rightarrow [B \rightarrow (A \wedge B)]$; t21, $A \rightarrow [B \rightarrow (A \vee B)]$ (equivalent to t20, $A \rightarrow (A \rightarrow A)$, given B_+) and t28, $A \rightarrow [B \rightarrow (C \rightarrow A)]$. But the addition of any of these schemes gives as a result strong systems. Ac-tually, in Proposition 7.7 the reader can find a wealth of extensions of B_c in which none of the referred schemes is provable. An obvious alternative is, of course, to add the rule Veq, $A \Rightarrow B \rightarrow A$, but t20, t21, and t25 are then immediate. And, on the other hand, as pointed out more than once above, the rule Veq is interpretable in RM_0-semantics and basic constructive logics built upon B_{K+} (i.e., B_+ plus Veq) will be treated in the following chap-ter. Thus, it seems that we need some appropriate restriction of Veq. Now, restriction of Veq to conditional theorems (i.e., $A \rightarrow B \Rightarrow C \rightarrow (A \rightarrow B)$) would not do, as it is equivalent to t25, given B_+. So, we propose restric-tion of Veq to negation theorems (i.e., $\neg A \Rightarrow \neg A \rightarrow (B \rightarrow \neg A)$), or, more precisely, VeqDn, which is equivalent to it, given B_+ plus A7 and Red. Consider the following proposition.

Proposition 2.40 (B'_{cS} and B_{cS} are equivalent). *Let B'_{cS} be the result of replacing in B_{cS} the rule VeqDn by the following rule: $\neg A \Rightarrow \neg A \rightarrow (B \rightarrow \neg A)$. Then, B'_{cS} and B_{cS} are deductively equivalent.*

Proof. (a) B'_{cS} includes B_{cS}. It is immediate by using T4 (provable with A1 and Red). (b) B_{cS} includes B'_{cS}. It is proved by using T3, T6, and T9.   □

The rule VeqDn has been chosen since (a) the rule Double negation (T4, $A \Rightarrow \neg\neg A$) is derivable in B_+ expanded with Red; and (b) the rule T5 ($A \Rightarrow B \rightarrow \neg\neg A$), immediate by T4 and VeqDn, guarantees that $\neg\neg A$ (A being a theorem) is present in any non-empty theory.

Concerning B_c, from the discussion above, it follows that it is the basic constructive logic in RM_1-semantics when $S = K$ in the same sense as B_{cS} is the basic constructive logic in RM_1-semantics when $S \subseteq K$. Finally, recall that, given B_+, A7 (respectively, A8), Red, and VeqDn are independent from each other.

A last remark. The study of the positive logics B_+t20 (i.e., B_+ plus t20) and B_+t25 (i.e., B_+ plus t25) can be carried out in RM_0-semantics. Intuitionistic-type negation can be introduced by simply adding A7 and Red (or A8 and Red), since VeqDn can be dropped in both systems as explained above. Models for these logics are provided by changing in B_{cS}-models (respectively, B_c-models) P4 (respectively, P5) by Pd (respectively, Pg) and adding the postulate pt20 (respectively, pt25) in the case of B_+t20 (respectively, B_+t25) plus A7 (or A8) and Red (cf. Chapter 6). Although we do not have space to treat them here, we remark that these systems are very interesting because a number of their extensions do not collapse in logics with the rule Veq (cf. Proposition 7.7 and Proposition 7.10). Thus, they are a, so to speak, bridge between the logics lacking paradoxes of relevance and those with the strongest of them, among which the rule Veq is generally to be found.

CHAPTER 3

The basic positive logic B_{K+}. The basic constructive logics B_{KS} and B_K

In this chapter, we study the logics B_{K+}, B_{KS}, and B_K. The logic B_{K+} is the basic positive logic in RM_0-semantics (RM-semantics without a set of designated points). Then, the logics B_{KS} and B_K are defined from B_{K+} as B_{cS} and B_c were defined from B_+ in the precedent chapter. Consequently (cf. §2.6), the logic B_{KS} is the basic constructive logic in RM_0-semantics with a set $S \subseteq K$ (RM-semantics without a set of designated points but with a set of consistent points); and the logic B_K is the basic constructive logic in RM_0-semantics with $S = K$ (RM-semantics in which all points are consistent but there is not a set of designated points). As pointed out above, unless otherwise stated, consistency is understood as w-consistency throughout this book (cf. the paragraph following Definition 0.8).

3.1 THE LOGIC B_{K+} AND ITS SEMANTICS

The logic B_{K+} is the result of adding the rule Veq ("Verum e quodlibet" —"A true proposition follows from any proposition"), also known as rule K, to the logic B_+ (the logic B_{K+} is introduced in [27] and [32]).

Definition 3.1 (The logic B_{K+}). The logic B_{K+} is axiomatized with A1–A6 and the rules MP, Adj, Suf, and Pref of B_+ (cf. Definition 1.10) and, in addition, with the rule Veq, or rule K:

$$K.\ A \Rightarrow B \to A$$

Proposition 3.2 (Some theorems of B_{K+}). *In addition to the theorems and rules of B_+ (cf. those remarked in Proposition 1.11), the following theorem and rule will be useful:*

$t7.\ B \to (A \to A)$	A1, K
$t8.\ A \to B \Rightarrow (A \to B) \to [C \to (A \to B)]$	K

Routley-Meyer Ternary Relational Semantics for Intuitionistic-Type Negations.
DOI: http://dx.doi.org/10.1016/B978-0-08-100751-8.00004-3

In what follows, B_{K+}-semantics is defined. As in the case of B_+, we shall first consider EB_{K_+}-models, models for strengthenings of B_{K+}, EB_{K+}-logics.

Definition 3.3 (EB_{K+}-models). An EB_{K+}-model M is a structure with at least the following items (1) A set K; (2) a ternary relation R defined on K subject to the following definitions and postulates for all $a, b, c, d \in K$:

$$d1.\ a \leq b =_{df} \exists x \in K\ Rxab$$
$$d2.\ R^2 abcd =_{df} \exists x \in K(Rabx\ \&\ Rxcd)$$
$$P1.\ a \leq a$$
$$P2.\ (a \leq b\ \&\ Rbcd) \Rightarrow Racd$$

(3) a (valuation) relation $\models$ from K to the set of all wffs $\mathcal{F}$ such that the following conditions (clauses) are satisfied for every $p \in \mathcal{P}$, $A, B \in \mathcal{F}$, and $a \in K$:

(i). $(a \leq b\ \&\ a \models p) \Rightarrow b \models p$

(ii). $a \models A \wedge B$ iff $a \models A$ and $a \models B$

(iii). $a \models A \vee B$ iff $a \models A$ or $a \models B$

(iv). $a \models A \to B$ iff for all $b, c \in K$, $(Rabc\ \&\ b \models A) \Rightarrow c \models B$

Additional elements of M are the following: (1) a set of postulates $Pj_1, ..., Pj_n$; (2) a non-empty subset S of K, and (3) the following conditions (clauses) for every $a \in K$ (cf. Definition 1.4):

(v). $a \models \neg A$ iff for all $b, c \in K$, $(Rabc\ \&\ c \in S) \Rightarrow b \not\models A$

(vi). $a \models \neg A$ iff for all $b, c \in K$, $Rabc \Rightarrow b \not\models A$

(vii). $a \models f$ iff $a \notin S$

(ix). $a \not\models f$

(x). $a \models t$

If clause (vii) is added, then the postulate Pf noted below is also added for all $a, b \in K$:

$$Pf.\ (a \leq b\ \&\ b \in S) \Rightarrow a \in S$$

Next, the accompanying notions of truth and validity are defined.

Definition 3.4 (Truth in a class of EB_{K+}-models). Let a class of EB_{K+}-models $\mathcal{M}$ be defined. A wff A is true in a model $M \in \mathcal{M}$ iff $a \vDash A$ for all $a \in K$ in M. Given an EB_{K+}-model M, $a \nvDash A$ symbolizes that A is not true in M.

Definition 3.5 (Validity in a class of EB_{K+}-models). Let a class of EB_{K+}-models $\mathcal{M}$ be defined. A wff A is $\mathcal{M}$-valid (in symbols $\vDash_{\mathcal{M}} A$) iff A is true in all $M \in \mathcal{M}$ ($\nvDash_{\mathcal{M}} A$ symbolizes that A is not $\mathcal{M}$-valid).

Subscripts are generally deleted when there is no danger of confusion. By the term EB_{K+}-validity we shall generally refer to validity in any class of EB_{K+}-models (cf. Definition 1.36).

Then, notice that EB_+-models (cf. Definition 1.4) and EB_{K+}-models are distinguished by the absence of a set of designated points in the latter where, consequently, validity is defined w.r.t. the set K of all points instead of w.r.t. the set of designated points as it is the case in EB_+-models. On the other hand, B_{K+}-models are the simplest EB_{K+}-models.

Definition 3.6 (B_{K+}-models). A B_{K+}-model is a basic EB_{K+}-model with no additional items added. That is, a B_{K+}-model is an EB_{K+}-model $(K, R, \vDash)$ where R is subject to d1, d2, P1, and P2, and $\vDash$ to just clauses (i)–(iv).

Below, we note a proposition recording some semantical postulates holding in all EB_{K+}-models (cf. Propositions 2.3 and 2.31). Then, we proceed to set the context in which soundness of EB_{K+}-logics can be proved.

Proposition 3.7 (Postulates holding in all EB_{K+}-models). *The following semantical postulates hold in all EB_{K+}-models:*

$$Ph.\ Rabc \Rightarrow b \leq c$$
$$Pi.\ R^2 abcd \Rightarrow Rbcd$$

Proof. Ph: immediate by d1; Pi: by d1, d2, and P2. □

The context in which soundness of EB_{K+}-logics is proved depends upon the Hereditary Condition and Entailment Lemmas, as it always obtains in RM-semantics.

Lemma 3.8 (Hereditary Condition). *For any EB_{K+}-model, $a, b \in K$ and $A \in \mathcal{F}$, $(a \leq b\ \&\ a \vDash A) \Rightarrow b \vDash A$.*

Proof. The proof is based upon clause (i) and P2 (cf. Lemma 1.8), not in the particular definition of the binary relation $\leq$. Consequently, the proof provided for Lemma 1.8 (Hereditary Condition for EB_+-models) is applicable here *mutatis mutandis*.   □

Lemma 3.9 (Entailment Lemma). *Let a class of EB_{K+}-models $\mathcal{M}$ be defined. For any $A, B \in \mathcal{F}$, $\vDash_{\mathcal{M}} A \to B$ iff $(a \vDash A \Rightarrow a \vDash B$ for all $a \in K)$ in all $M \in \mathcal{M}$.*

Proof. The proof is not based on the particular choice of the set (either K or else O) upon which validity is defined, but upon d1 and P1, which work similarly regardless of which set has finally been chosen. In consequence, the proof given for Lemma 1.9 (Entailment Lemma for EB_+-models) would here hold *mutatis mutandis*.   □

We turn to the proof that all theorems of B_{K+} are valid in all EB_{K+}-models (cf. Definition 1.36), whence soundness of B_{K+} will be derived as a corollary.

Proposition 3.10 (All theorems of B_{K+} are EB_{K+}-valid). *For any wff A, if $\vdash_{B_{K+}} A$, then A is EB_{K+}-valid.*

Proof. The proof of A1–A6, Adj, Suf, and Pref is based on the Entailment Lemma and clauses (i)–(iv) (cf. Proposition 1.13). Therefore, regarding these axioms and rules, the proof of Proposition 3.10 is similar to the proof of the corresponding proposition for EB_+-models, Proposition 1.13. So, let us prove that MP and K preserve EB_{K+}-validity.

(a) *MP, A & $A \to B \Rightarrow B$, preserves EB_{K+}-validity*: (Cf. Proposition 1.13.) Let $a \in K$ in an arbitrary EB_{K+}-model M and suppose for $A, B \in \mathcal{F}$, (1) $\vDash A$ and (2) $\vDash A \to B$. We have to prove $a \vDash B$. By d1 and P1, we have (3) $Rxaa$ for some $x \in K$ in M. Then, we have (4) $a \vDash A$ (by 1 and Definition 3.5) and (5) $x \vDash A \to B$ (by 2 and Definition 3.5). Finally, (6) $a \vDash B$ is derivable (by 3, 4, 5, and iv), as it was to be proved.

(b) *K, $A \Rightarrow B \to A$, preserves EB_{K+}-validity*: Suppose that there are $A, B \in \mathcal{F}$ such that (1) $\vDash A$ and (2) $\nvDash B \to A$. By Lemma 3.9, there is $a \in K$ in some EB_{K+}-model such that (3) $a \vDash B$ and (4) $a \nvDash A$. But 1 and 4 contradict each other by Definition 3.5.   □

Corollary 3.11 (Soundness of B_{K+}). *For any $A \in \mathcal{F}$, if $\vdash_{B_{K+}} A$, then $\vDash_{B_{K+}} A$.*

Proof. Immediate by Proposition 3.10.   □

3.2 COMPLETENESS OF B_{K+}

In this section, we proceed to set the preliminaries to the completeness proofs of EB_{K+}-logics. In the rest of this section, L has to be considered as an EB_{K+}-logic and members in K^T, K^P, etc. have to be understood as EB_{K+}-theories, not merely as EB_+-theories (cf. Definitions 1.14, 1.15, and 1.36).

As we have viewed in the completeness proofs proved so far, canonical worlds or canonical points are theories with certain properties. Concerning EB_{K+}-semantics, we have just seen that validity is defined w.r.t. the set of all points. Consequently, we have to correspondingly demand that all theories be regular in canonical EB_{K+}-models, which is in fact a distinctive feature of this type of canonical models. This of course means that each new theory introduced in the development of the completeness proof has to be shown regular. But Lemma 3.12 below shows that an EB_{K+}-theory is regular iff it is non-empty, whence from now on it will be sufficient to prove that EB_{K+}-theories have the latter property. After proving Lemma 3.12, we examine whether the propositions and lemmas on which the completeness of B_+ is based are still applicable when we move to B_{K+}.

Lemma 3.12 (Regularity and non-emptiness). *Let $a \in K^T$. a is regular iff a is non-empty.*

Proof. (a) ($\Rightarrow$) It is obvious. (b) ($\Leftarrow$) Let A be a theorem of L and $B \in a$. By the rule K, $\vdash_L B \to A$ follows. So, we have $A \in a$. $\qquad\square$

Turning now to the propositions and lemmas used in the completeness proof of B_+, we remark that Lemmas 1.17, 1.18, and 1.19 are still adequate for EB_{K+}-logics inasmuch as the new theories built up in each one of these lemmas, from their respective hypotheses, are non-empty if the theories they extend are also non-empty. However, Lemma 1.22 and Corollary 1.23 (along with Definition 1.20 and 1.21) have to be —slightly— modified as the binary relation $\leq$ is defined w.r.t. the set K in EB_{K+}-models, not w.r.t. O as in EB_+-models. In particular, we have:

Definition 3.13 ($a \leq^T b$). Let $a, b \in K^T$. Then, $a \leq^T b =_{df} \exists x \in K^{NT} R^T xab$.

Definition 3.14 ($a \leq^P b$). Let $a, b \in K^P$. Then, $a \leq^P b =_{df} \exists x \in K^{NP} R^P xab$.

But given Lemma 3.12, Lemma 3.15, and Corollary 3.16 below are in fact equivalent to Lemma 1.22 and Corollary 1.23 and are proved similarly.

Lemma 3.15 ($a \leq^T b$ iff $a \subseteq b$). *For any $a, b \in K^T$, $a \leq^T b$ iff $a \subseteq b$.*

Corollary 3.16 ($a \leq^P b$ iff $a \subseteq b$). *For any $a, b \in K^P$, $a \leq^P b$ iff $a \subseteq b$.*

Thus, once the following important lemma is proved, the completeness proof of B_{K_+} mirrors that of B_+.

Lemma 3.17 (R^T and non-emptiness). *Let $a, b \in K^{NT}$, $c \in K^T$, and $R^T abc$. Then $c \in K^{NT}$.*

Proof. Let $B \in a$, $A \in b$, and $R^T abc$. By t7, we have $\vdash_{B_{K_+}} B \to (A \to A)$. So, $A \to A \in a$ follows and then $A \in c$. $\qquad\qquad\qquad\qquad\qquad\qquad\square$

Lemma 3.17 (for EB_{K_+}-logics) has the same content as Lemma 2.15 (for EB_{cS}-logics). But the latter is proved by using the rule VeqDn whereas the former is proved by t7, as we have just seen. As noted in Remark 2.16 and discussed in §2.6, Lemma 2.15 (or Lemma 3.17) is provable when a theorem of the form $A \to (B \to C)$ ($A \neq B$) is at our disposal. Lemma 3.17 (and Lemma 2.15) are essential for proving theories non-empty in the completeness proofs of EB_{K_+}-logics (respectively, EB_{cS}-logics), particularly in the proof of the canonical adequacy of certain semantical postulates.

Next, canonical EB_{K_+}-models are defined and the completeness of B_{K_+} is established.

Definition 3.18 (Canonical EB_{K_+}-models). There are three possible types of canonical models for EB_{K_+}-logics: (a) $(K^{NP}, R^{NP}, \vDash^{NP})$; (b) $(K^{NP}, S^{NP}, R^{NP}, \vDash^{NP})$; and (c) $(K^{WNP}, R^{WNP}, \vDash^{WNP})$ where superscripts W, N, and P are read as in Definition 1.26.

Definition 3.19 (The canonical B_{K_+}-model). The canonical B_{K_+}-model is a canonical EB_{K_+}-model of type a. That is, it is the structure $(K^{NP}, R^{NP}, \vDash^{NP})$, where K^{NP}, R^{NP}, and $\vDash^{NP}$ are referred to B_{K_+}-theories.

Lemma 3.20 (The canonical model is a model). *The canonical B_{K_+}-model is indeed a B_{K_+}-model.*

Proof. It suffices to prove the facts (1)–(3) listed below.

(1) The set K^{NP} is not empty: by Lemma 1.17, as B_{K_+} is a regular and non-trivial B_{K_+}-theory.

(2) Postulates P1 and P2 hold in the canonical B_{K_+}-model: immediate by Corollary 3.16.

(3) Clauses (i)–(iv) in Definition 3.6 are satisfied by the canonical B_{K_+}-model: clause (i), by Lemma 1.25a and Corollary 3.16; clauses (ii)

and (iii), by Lemma 1.25b and Lemma 1.25c, respectively; clause (iv) (from left to right) is proved by Lemma 1.25d; finally, clause (iv) (from right to left) is proved by Lemma 1.25e and Lemma 3.17 that guarantees that the theory y obtained in Lemma 1.25e is actually non–empty.   □

Theorem 3.21 (Completeness of B_{K+}). *For any $A \in \mathcal{F}$, if $\vDash_{B_{K+}} A$, then $\vdash_{B_{K+}} A$.*

Proof. By Lemmas 1.17 and 3.20 and Definitions 3.6 and 3.19 (cf. Remark 1.31).   □

3.3 THE LOGIC B_{KS}

The logic B_{KS} is built up from B_{K+} similarly as B_{cS} was defined from B_+. It is built upon the positive language of B_{K+} expanded with the unary connective ¬ (negation). B_{KS} is the basic constructive logic in the RM-semantics without a set of designated points, but with a set of consistent points, S.

Definition 3.22 (The logic B_{KS}). The logic B_{KS} is axiomatized by adding the following axioms to B_{K+}:

$$A9.\ \neg A \to [A \to \neg(A \to A)]$$
$$A10.\ [B \to \neg(A \to A)] \to \neg B$$

We remark some theorems and derived rules of B_{KS} (a proof is sketched for each one of them):

T1. $A \to B \Rightarrow \neg B \to \neg A$	A9, A10
T2. $A \Rightarrow \neg A \to \neg B$	K, T1
T3. $A \Rightarrow \neg(B \to B) \leftrightarrow \neg A$	T2
T4. $B \Rightarrow (A \to \neg B) \to \neg A$	T3, A10
T5. $\neg A \to [A \to \neg(B \to B)]$	K, A9, T1

T6. $A \Rightarrow \neg\neg A \to (B \to \neg\neg A)$

Proof sketch.

1. $A \,\&\, C$ Hyp

$$2.\ C \to A \hspace{6cm} \text{K, 1}$$

$$3.\ \neg A \to \neg C \hspace{5.5cm} \text{T1, 2}$$

$$4.\ (\neg A \to \neg C) \to [B \to (\neg A \to \neg C)] \hspace{2cm} \text{t8, 3}$$

$$5.\ \neg\neg A \to (\neg A \to \neg C) \hspace{4cm} \text{T3, T5, 1}$$

$$6.\ (\neg A \to \neg C) \to \neg\neg A \hspace{4.3cm} \text{T4, 1}$$

$$7.\ \neg\neg A \to (B \to \neg\neg A) \hspace{4.2cm} \text{4, 5, 6}$$

$$\square$$

Thus, we see, B_{cS} is a sublogic of B_{KS}.

Proposition 3.23 (B_{cS} is included in B_{KS}). *The logic B_{cS} is included in B_{KS}, that is, for any $A \in \mathcal{F}$, if $\vdash_{B_{cS}} A$, then $\vdash_{B_{KS}} A$.*

Proof. Immediate as B_{cS} is axiomatized by adding T4, T5, and T6 to B_+ (cf. Definition 2.1). $\hspace{8cm} \square$

We now turn to the definition of an RM-semantics for B_{KS}.

Definition 3.24 (B_{KS}-models). A B_{KS}-model is an EB_{K+}-model (K, S, R, $\vDash$) where clauses (i)–(v) and the following semantical postulates (in addition to P1 and P2) are satisfied for all $a, b, c \in K$:

$$\text{P3. } (Rabc\ \&\ c \in S) \Rightarrow a \in S$$

$$\text{P6. } (Rabc\ \&\ c \in S) \Rightarrow \exists x \in K\ \exists y \in S\ Rcxy$$

We proceed to the proof of soundness of B_{KS} following the pattern set for the systems previously studied, B_+, B_{cS}, and B_c (cf. Definition 1.36 on the notions of an EB_{KS}-model and EB_{KS}-validity).

Proposition 3.25 (All theorems of B_{KS} are EB_{KS}-valid). *For any $A \in \mathcal{F}$, if $\vdash_{B_{KS}} A$, then A is EB_{KS}-valid.*

Proof. We prove the EB_{KS}-validity of A9 and A10 as A1–A6, MP, Adj, Suf, Pref, and K are taken care of in Proposition 3.10. Now, A9 is proved similarly as A7 of B_{cS} (cf. Proposition 2.4 —notice that the postulate P3 holds in both EB_{cS}-models and EB_{KS}-models). So, we prove the EB_{KS}-validity of A10.

(a) A10, $[B \to \neg(A \to A)] \to \neg B$, is EB_{KS}-valid: if A10 is not EB_{KS}-valid, by Lemma 3.9, there are $A, B \in \mathcal{F}$ and $a \in K$ in some EB_{KS}-model such that (1) $a \vDash B \to \neg(A \to A)$ and (2) $a \nvDash \neg B$. Then, by

2 and v, there are $b, c \in K$ such that (3) $Rabc$, (4) $c \in S$, and (5) $b \vDash B$. By 1, 3, 5, and iv, we have (6) $c \vDash \neg(A \to A)$. But by 3, 4, and P6, we get (7) $Rcxy$ with $x \in K$ and $y \in S$. Then, by 6, 7, and v, (8) $x \nvDash A \to A$ follows, which is impossible.   □

Corollary 3.26 (Soundness of B_{KS}). *For any $A \in \mathcal{F}$, if $\vdash_{B_{KS}} A$, then $\vDash_{B_{KS}} A$.*

Proof. Immediate by Proposition 3.25.   □

Bending to completeness, we comment that facts proved in Sections 2.3, 2.4, and 3.2 suffice to prove it. (Notice that any B_{KS}-theory is a B_{cS}-theory as B_{cS} is a sublogic of B_{KS}.) We set:

Definition 3.27 (The canonical B_{KS}-model). The canonical B_{KS}-model is a canonical EB_{K+}-model of type b, that is, it is the structure $(K^{NP}, S^{NP}, R^{NP}, \vDash^{NP})$ (cf. Definition 3.18).

Then, we prove:

Lemma 3.28 (The canonical model is a model). *The canonical B_{KS}-model is indeed a B_{KS}-model.*

Proof. (1) The set S^{NP} is not empty: By Lemma 2.13, given that B_{KS} is a w-consistent, regular B_{KS}-theory.

(2) Postulates P1–P3 and P6 hold in the canonical B_{KS}-model: P1 and P2, by Corollary 3.16; P3 by Lemma 2.17; and P6, by Lemma 2.18(b).

(3) Clauses (i)–(v) hold in the canonical B_{KS}-model: Clauses (i)–(iv) are proved as in the case of the canonical B_{K_+}-model (Lemma 3.20). Then, clause (v) is proved as follows. (a) ($\Rightarrow$) Suppose $A \in \mathcal{F}$ and $a, b \in K^{NP}$, $\neg A \in a$, $R^{NP}abc$, and $c \in S^{NP}$. Then, $A \notin b$ follows by Lemma 2.21. (b) ($\Leftarrow$) Suppose that there are $A \in \mathcal{F}$ and $a \in K^{NP}$ such that $\neg A \notin a$. By Lemma 2.22, there are $x \in K^{NP}$ and $y \in S^P$ such that $R^P axy$ and $A \in x$. Then, $y \in S^{NP}$ follows by Lemma 3.17. Thus, we have $x \in K^{NP}$ and $y \in S^{NP}$ such that $R^{NP}axy$ and $A \in x$, as it was required.   □

Finally, we have completeness.

Theorem 3.29 (Completeness of B_{KS}). *For any $A \in \mathcal{F}$, if $\vDash_{B_{KS}} A$, then $\vdash_{B_{KS}} A$.*

Proof. By Lemmas 2.13 and 3.28 and Definitions 3.24 and 3.27 (cf. Remark 1.31).   □

3.4 THE LOGIC B_K

The logic B_K is defined from B_{K+} as B_c was defined from B_+. The logic B_K is the basic constructive logic in RM-semantics with neither a set of designated points nor a selected subset of consistent points. It is built upon the positive language of B_{K+} expanded with the unary connective $\neg$ (negation).

Definition 3.30 (The logic B_K). The logic B_K is axiomatized by adding the following axioms to B_{K+}:

$$\text{A8. } \neg A \to (A \to B)$$
$$\text{A10. } [B \to \neg(A \to A)] \to \neg B$$

Obviously, we have:

Proposition 3.31 (B_{KS} is included in B_K). *The logic B_{KS} is included in B_K. That is, for any $A \in \mathcal{F}$, if $\vdash_{B_{KS}} A$, then $\vdash_{B_K} A$.*

Proof. Immediate: B_{KS} is axiomatized by adding A9 $\neg A \to [A \to \neg(A \to A)]$ and A10 to B_{K+}. But A9 is an instance of A8. □

The relationship the logics studied so far maintain to each other are summarized in the following diagram (the arrow stands for set inclusion)

Then, notice that theorems of B_+, B_{K+}, B_{cS}, B_c, and B_{KS} remarked above are also theorems of B_K.

We now proceed to the definition of an RM-semantics for B_K.

Definition 3.32 (B_K-models). A B_K-model is an EB_{K+}-model $(K, R, \vDash)$ where clauses (i)–(iv) and (vi) and the following semantical postulate (in addition to P1 and P2) are satisfied:

$$\text{P7. } Rabc \Rightarrow \exists x \in K \ \exists y \in K \ Rcxy$$

Then, we prove (cf. Definition 1.36):

Proposition 3.33 (All theorems of B_K are EB_K-valid). *For any $A \in \mathcal{F}$, if $\vdash_{B_K} A$, then A is EB_K-valid.*

Proof. Given Proposition 3.10 (all theorems of B$_{K+}$ are EB$_{K+}$-valid), we have to consider only A8 and A10. But A8 is proved similarly as in EB$_c$-models (cf. Proposition 2.32) and the proof of A10 (similar to the one given above in Proposition 3.25) is as follows.

(a) *A10*, $[B \rightarrow \neg(A \rightarrow A)] \rightarrow \neg B$, *is EB$_K$-valid:* Suppose that there are $A, B \in \mathcal{F}$ and $a \in K$ in some EB$_K$-model such that (1) $a \vDash B \rightarrow \neg(A \rightarrow A)$ but (2) $a \nvDash \neg B$. By 2 and vi, there are $b, c \in K$ such that (3) $Rabc$ and (4) $b \vDash B$. By 1, 3, 4, and iv, we have (5) $c \vDash \neg(A \rightarrow A)$. But by 3 and P7 we get (6) $Rcxy$ for $x, y \in K$. Then, by 5, 6, and vi $x \nvDash A \rightarrow A$ is derivable, which is impossible. □

Soundness of B$_K$ is a corollary of Proposition 3.33.

Corollary 3.34 (Soundness of B$_K$). *For any $A \in \mathcal{F}$, if $\vdash_{B_K} A$, then $\vDash_{B_K} A$.*

Proof. Immediate by Proposition 3.33. □

As in the case of B$_{KS}$, facts proved in preceding sections will suffice to show that B$_K$ is complete.

Definition 3.35 (The canonical B$_K$-model). The canonical B$_K$-model is a canonical EB$_{K+}$-model of type c, that is, it is the structure (K^{WNP}, R^{WNP}, $\vDash^{WNP}$) (cf. Definition 3.18).

Lemma 3.36 (The canonical model is a model). *The canonical B$_K$-model is in fact a B$_K$-model.*

Proof. It is similar to that of Lemma 3.28 (notice that, by Lemma 2.17, if $R^{NT}abc$ for $a, b \in K^{NT}$ and $c \in K^{WNT}$, then $a, b \in K^{WNT}$). □

Finally, completeness is proved in the customary way.

Theorem 3.37 (Completeness of B$_K$). *For any $A \in \mathcal{F}$, if $\vDash_{B_K} A$, then $\vdash_{B_K} A$.*

Proof. By Lemmas 2.13 and 3.36 and Definitions 3.32 and 3.35 (cf. Remark 1.31). □

In §1.5, it was explained how to prove a completeness theorem (of sorts) for EB$_+$-logics interpreted with EB$_+$-semantics, i.e., RM-semantics with a set of designated points. In order to define a similar theorem for EB$_{K+}$-logics modelled with EB$_{K+}$-semantics, i.e., RM-semantics without a set of designated points O, the notion of "proof from premises" introduced in Definition 1.32 can still be maintained, but, of course, the notion of "semantical consequence relation" has to be modified being referred to the set K, not to the set O. Then, it should be read as follows.

Definition 3.38 (Semantical consequence relation). Let L, Γ, and A be as in Definition 1.32 and suppose given an RM-semantics for L built upon EB_{K+}-models (cf. Definitions 3.3 and 3.5). Then, $\Gamma \vDash_L A$ ("A is a semantical consequence of Γ in L-semantics") iff for any L-model M and $a \in K$, if $a \vDash \Gamma$, then $a \vDash A$ (for any model M and $a \in K$, $a \vDash \Gamma$ iff $a \vDash B$ for all $B \in \Gamma$).

Then, we have the following theorem that can be proved similarly as Theorems 1.34 and 1.35 in the case of EB_+-logics interpreted with EB_+-semantics.

Theorem 3.39 (Coextensiveness of $\vdash_L$ and $\vDash_L$). *For any set of wffs Γ and wff A, $\Gamma \vdash_L A$ iff $\Gamma \vDash_L A$.*

PART 2

CHAPTER 4

Logics definitionally equivalent to the basic constructive logics. The logics B_{cSf}, B_{cf}, B_{KSf}, and B_{Kf}

We provide logics (defined with the propositional falsity constant) definitionally equivalent to the basic logics B_{cS}, B_c, B_{KS}, and B_K defined in Chapters 2 and 3. But the first section of the chapter introduces a simple expansion with the constant f of each one of the logics B_+ and B_{K+}, the logics $B_{+,f}$ and $B_{K+,f}$. Then, the logics B_{cSf} and B_{cf} are extensions of $B_{+,f}$, whereas B_{KSf} and B_{Kf} are extensions of $B_{K+,f}$.

4.1 THE LOGICS $B_{+,f}$ AND $B_{K+,f}$

The logics $B_{+,f}$ and $B_{K+,f}$ are defined by adding the propositional falsity constant f to the positive language of both B_+ and B_{K+} (cf. Definition 1.1). No new axioms, however, are added, but negation can be introduced via the definition

$$D\neg. \quad \neg A =_{df} A \to f$$

In this way, some negation theorems and rules such as the ones that follow can be proved (cf. Proposition 1.11).

T1. $A \to B \Rightarrow \neg B \to \neg A$	Suf, D¬
T2. $\neg B \Rightarrow (A \to B) \to \neg A$	Pref, D¬
T3. $\neg(A \vee B) \leftrightarrow (\neg A \wedge \neg B)$	t6, D¬
T4. $(\neg A \vee \neg B) \to \neg(A \wedge B)$	t3, D¬
T5. $\neg f$	A1, D¬

Models are defined as follows:

Definition 4.1 ($B_{+,f}$-models; $B_{K+,f}$-models). A $B_{+,f}$-model is an EB_+-model $(K, O, S, R, \vDash)$ and a $B_{K+,f}$-model is an EB_{K+}-model $(K, S, R, \vDash)$

Routley-Meyer Ternary Relational Semantics for Intuitionistic-Type Negations.
DOI: http://dx.doi.org/10.1016/B978-0-08-100751-8.00006-7

where clauses (i)–(iv), (vii) and postulates P1, P2, and Pf hold. (We recall that Pf is the following postulate: $(a \leq b\ \&\ b \in S) \Rightarrow a \in S$. Cf. Definition 1.4, Definition 3.3.)

We remark that f is not $EB_{+,f}$-valid ($EB_{K+,f}$-valid) (cf. Definition 1.36).

Proposition 4.2 (f is neither $EB_{+,f}$-valid nor $EB_{K+,f}$-valid). *The constant f is not $EB_{+,f}$-valid (respectively, $EB_{K+,f}$-valid). In fact, f is false in each $a \in O \cap S$ (respectively, $a \in S$) in each $EB_{+,f}$-model (respectively, $EB_{K+,f}$-model).*

Proof. Let M be any $EB_{+,f}$-model (respectively, $EB_{K+,f}$-model), and $a \in O \cap S$ (respectively, $a \in S$). Then, we have $a \nvDash f$ by clause (vii). (Recall that $O \cap S$ —respectively, S— is not empty; cf. Definitions 1.4 and 3.3).   □

On the other hand, as immediate consequences of Propositions 1.13 and 3.10, we have the following proposition and corollary. (Notice that Lemmas 1.8 and 1.9 —respectively, 3.8 and 3.9— hold for $EB_{+,f}$-logics —respectively, $EB_{K+,f}$-logics.)

Proposition 4.3 ($EB_{+,f}$ ($EB_{K+,f}$)-validity of theorems). *For any $A \in \mathcal{F}$, (a) if $\vdash_{B_{+,f}} A$, then A is $EB_{+,f}$-valid; (b) if $\vdash_{B_{K+,f}} A$, then A is $EB_{K+,f}$-valid.*

Proof. Immediate. (a) By Proposition 1.13; (b) by Proposition 3.10.   □

Corollary 4.4 (Soundness of $B_{+,f}$ and $B_{K+,f}$). *For any $A \in \mathcal{F}$, (a) if $\vdash_{B_{+,f}} A$, then $\vDash_{B_{+,f}} A$; (b) if $\vdash_{B_{K+,f}} A$, then $\vDash_{B_{K+,f}} A$.*

Proof. Immediate. (a) By Proposition 4.3(a); (b) by Proposition 4.3(b).   □

Turning to completeness, let L be an $EB_{+,f}$-logic and recall that U^T is the set of all u-consistent L-theories (cf. Definitions 1.16 and 1.26). We prove a couple of preliminary propositions.

Proposition 4.5 ($a \in U^T$ iff $f \notin a$). *Let $a \in K^T$. Then, $a \in U^T$ iff $f \notin a$.*

Proof. (a) ($\Rightarrow$) Suppose $a \in U^T$. If $f \in a$, then, by T5 ($\neg f$), a contains the argument of a negative theorem contradicting the u-consistency of a. (b) ($\Leftarrow$) Suppose $f \notin a$. If $\neg A$ (i.e., $A \to f$) is a theorem of L, then $A \notin a$ follows: otherwise $f \in a$ (a is closed by L-imp).   □

Proposition 4.6 ($B_{+,f}$ and $B_{K+,f}$ are u-consistent). *The logics $B_{+,f}$ and $B_{K+,f}$ are u-consistent.*

Proof. By Proposition 4.2 f is not $B_{K+,f}$-valid. Then, by the soundness theorem (Corollary 4.4), it follows that f is not derivable in $B_{K+,f}$. Thus, $B_{K+,f}$ is u-consistent by Proposition 4.5, whence $B_{+,f}$ is u-consistent as well, since $B_{+,f}$ is included in $B_{K+,f}$. $\qquad\square$

Next, we proceed to the definition of the canonical models and to the completeness proofs.

Definition 4.7 (The canonical $B_{+,f}$-model and $B_{K+,f}$-model). The canonical $B_{+,f}$-model is the structure $(K^P, O^P, U^P, R^P, \vDash^P)$ and the canonical $B_{K+,f}$-model is the structure $(K^{NP}, U^{NP}, R^{NP}, \vDash^{NP})$ (cf. Definition 1.26).

Notice that the set S in Definition 4.1 is interpreted as U^P and U^{NP} in the canonical $B_{+,f}$-model and the canonical $B_{K+,f}$-model, respectively.

Lemma 4.8 (The canonical model is a model). *(a) The canonical $B_{+,f}$-model is in fact a $B_{+,f}$-model; (b) the canonical $B_{K+,f}$-model is in fact a $B_{K+,f}$-model.*

Proof. (a) Postulates P1, P2, and Pf are immediate by Corollary 1.23, and clauses (i)–(iv) follow by Lemma 1.25. Then, $O^P \cap U^P$ is not empty by Proposition 4.6 and Lemma 1.17, and clause (vii) holds canonically by Proposition 4.5. (b) Postulates P1, P2, and Pf are immediate by Corollary 3.16, and clauses (i)–(iv) are proved as in Lemma 3.20. Finally, that U^{NP} is not empty and that clause (vii) holds canonically is proved similarly as in the case of $B_{+,f}$. $\qquad\square$

Theorem 4.9 (Completeness of $B_{+,f}$ and $B_{K+,f}$). *For any $A \in \mathcal{F}$, (a) if $\vDash_{B_{+,f}} A$, then $\vdash_{B_{+,f}} A$; (b) if $\vDash_{B_{K+,f}} A$, then $\vdash_{B_{K+,f}} A$.*

Proof. (a) By Definitions 4.1 and 4.7, Lemmas 1.17 and 4.8, and Proposition 4.6; (b) the proof is similar (cf. Remark 1.31). $\qquad\square$

We remark that expansions of B_+ and B_{K+} with f when this constant is interpreted according to clause (ix) require the thesis $f \rightarrow A$ (labelled A8$_f$ in the following section) and are treated in §6.3.

4.2 THE LOGICS B_{cSf}, B_{cf}, B_{KSf}, AND B_{Kf}

The logics B_{cSf}, B_{cf}, B_{KSf}, and B_{Kf} are definitionally equivalent to B_{cS}, B_c, B_{KS}, and B_K, respectively. By ¬-logics, we shall generally refer to the latter logics and by f-logics to the former. In each one of these logics, negation can be introduced (as in $B_{+,f}$ and $B_{K,f}$) according to the definition D¬: $\neg A =_{df} A \to f$. On the other hand, in each one of the basic constructive logics, f can be introduced via the definition Df: $f =_{df} \neg(A \to A)$. The basic constructive f-logics can be defined as follows.

The logic B_{cSf} is an extension of $B_{+,f}$.

Definition 4.10 (The logic B_{cSf}). The logic B_{cSf} is axiomatized by adding the following rules to $B_{+,f}$:

$$A7_f.\ A \Rightarrow f \to (A \to f)$$
$$Red_f.\ A \Rightarrow (A \to f) \to f$$
$$VeqDn_f.\ A \Rightarrow [(A \to f) \to f] \to [B \to [(A \to f) \to f]]$$

The logic B_{cf} is an extension of B_{cSf}.

Definition 4.11 (The logic B_{cf}). The logic B_{cf} is axiomatized by adding the following rules and axiom to $B_{+,f}$: Red_f, $VeqDn_f$, and

$$A8_f.\ f \to A$$

instead of $A7_f$.

The logics B_{KSf} and B_{Kf} are extensions of $B_{K+,f}$, the latter logic being in its turn an extension of the former.

Definition 4.12 (The logic B_{KSf}). The logic B_{KSf} is axiomatized by adding $A7_f$ and Red_f to $B_{K+,f}$.

Definition 4.13 (The logic B_{Kf}). The logic B_{Kf} is axiomatized by adding $A8_f$ and Red_f to $B_{K+,f}$.

4.3 DEFINITIONAL EQUIVALENCE

We now proceed to prove the definitional equivalence of each f-logic with its corresponding ¬-logic. Firstly, we prove the definitional equivalence of B_{cS} and B_{cSf}.

We shall understand the notion of definitional equivalence as "definitional equivalence via translations" (see, e.g., [4]). For our present purposes, this notion can be explained as follows. Let L1 and L2 be two logics in different languages, t1 the set of terms of L1 absent in L2, and t2, the set of terms of L2 absent in L1. Then L1 and L2 are definitionally equivalent iff there are definitions of t1 in terms of L2 (Dt1) and definitions of t2 in terms of L1 (Dt2) such that L1 ∪ {Dt2} = L2 ∪ {Dt1}. ($x \cup y$ is the deductive closure of the union of x and y, and definitions are expressed as a set of suitable biconditionals.) It is important to note that it is not sufficient to prove L1 ⊆ L2 ∪ {Dt1} and L2 ⊆ L1 ∪ {Dt2}. It additionally has to be shown that Dt2 is provable in L2 ∪ {Dt1} and Dt1 is provable in L1 ∪ {Dt2} (cf. [4]).

Therefore, we have to prove in the present case:

1. $B_{cs,f} \subseteq B_{cS} \cup \{Df\}$
2. $B_{cS} \subseteq B_{cSf} \cup \{D\neg\}$
3. $D\neg$ is provable in $B_{cS} \cup \{Df\}$
4. Df is provable in $B_{cSf} \cup \{D\neg\}$

where D¬ is $\neg A =_{df} A \to f$ and Df is $f =_{df} \neg(A \to A)$. That is, $\neg A$ replaces the formula $A \to f$ and f replaces any wff of the form $\neg(A \to A)$. Recall that the following is provable in B_{cS}: $A \mathrel{\&} B \Rightarrow \neg A \leftrightarrow \neg B$ (T13). Consequently in any EB_{cS}-logic L, we have: for any $A, B \in \mathcal{F}$, $\neg(A \to A)$, and $\neg(B \to B)$ are equivalent, and so the definition of f is legitimate in L since the *definiens* formula does not depend on the choice of A.

Next, we turn to the proof of the definitional equivalence according to the scheme presented above. Let L be any EB_{cS}-logic and L′ any EB_{cSf}-logic.

Proposition 4.14 (D¬ is provable in $L \cup \{Df\}$). *Definition D¬ is provable in the logic L plus Df.*

Proof. D¬ interpreted with Df is (1) $\neg A \leftrightarrow [A \to \neg(A \to A)]$. Then 1 is immediate by A7 and T1 of B_{cS}. □

Proposition 4.15 (Df is provable in $L' \cup \{D\neg\}$). *Definition Df is provable in the logic L′ plus D¬.*

Proof. By A7$_f$ and Red$_f$, we have $f \leftrightarrow [(A \to A) \to f]$ whence Df, $f \leftrightarrow \neg(A \to A)$, follows by D¬. □

Proposition 4.16 ($B_{cSf} \subseteq B_{cS} \cup \{Df\}$). *All axioms and rules of B_{cSf} are provable in B_{cS} plus Df.*

Proof. It suffices to prove A7$_f$, Red$_f$, and VeqDn$_f$.

(a) A7$_f$, $A \Rightarrow f \to (A \to f)$: By T11, we have (1) $A \Rightarrow \neg A \to [A \to \neg(A \to A)]$, whence by T13 and B$_+$, we get (2) $A \Rightarrow \neg(A \to A) \to [A \to \neg(A \to A)]$. Finally, A7$_f$ follows from 2 by Df.

(b) Red$_f$, $A \Rightarrow (A \to f) \to f$: By T14, we have $A \Rightarrow [A \to \neg(A \to A)] \to \neg(A \to A)$, whence Red$_f$ follows by Df.

(c) VeqDn$_f$, $A \Rightarrow [(A \to f) \to f] \to [B \to [(A \to f) \to f]]$, is immediate by T16 and and Df.   □

Proposition 4.17 (B$_{cS}$ $\subseteq$ B$_{cSf}$ $\cup$ {D¬}). *All axioms and rules of B$_{cS}$ are provable in B$_{cSf}$ plus D¬.*

Proof. (a) A7, $\neg A \to [A \to \neg(B \to B)]$: By A7$_f$, we have (1) $f \to [(B \to B) \to f]$, whence (2) $(A \to f) \to [A \to [(B \to B) \to f]]$ follows by Pref. Finally, we have A7 by 2 and D¬.

(b) Red, $A \Rightarrow (B \to \neg A) \to \neg B$: By Red$_f$ and Pref we have $A \Rightarrow [B \to (A \to f)] \to (B \to f)$, whence we get Red by D¬.

(c) The rule VeqDn, $A \Rightarrow \neg\neg A \to (B \to \neg\neg A)$, is immediate by VeqDn$_f$ and D¬.   □

Theorem 4.18 (B$_{cS}$ and B$_{cSf}$ are definitionally equivalent). *The logics B$_{cS}$ and B$_{cSf}$ are definitionally equivalent.*

Proof. By Propositions 4.14, 4.15, 4.16, and 4.17.   □

The definitional equivalence of the rest of the f-logics with their respective ¬-logics is a consequence of the propositions proved above, since B$_{cf}$, B$_{KSf}$, and B$_{Kf}$ are extensions of B$_{cSf}$. In particular, given Propositions 4.14–4.17, the definitional equivalence theorems are proved as follows.

Theorem 4.19 (B$_c$ and B$_{cf}$ are definitionally equivalent). *B$_c$ and B$_{cf}$ are definitionally equivalent.*

Proof. It suffices to prove the facts 1 and 2 below.

(1) The axiom A8$_f$ is a theorem of B$_c$ plus Df: By A8, we have $\neg\neg(B \to B) \to [\neg(B \to B) \to A]$; and by A1 and T4, $\neg\neg(B \to B)$. Then, we get $\neg(B \to B) \to A$, whence A8$_f$ is immediate by Df.

(2) The axiom A8 is a theorem of B$_{cf}$ plus D¬: By A8$_f$ and Pref, we get $(A \to f) \to (A \to B)$, hence $\neg A \to (A \to B)$ follows by D¬.   □

Theorem 4.20 (B$_{KS}$ and B$_{KSf}$ are definitionally equivalent). *The logics B$_{KS}$ and B$_{KSf}$ are definitionally equivalent.*

Proof. It suffices to prove the facts 1 and 2 below.

(1) All axioms and rules of B$_{KS}$ are provable in B$_{KSf}$ plus D¬: By Proposition 4.17, A7 and Red are provable from B$_+$ plus A7$_f$ and Red$_f$.

But A9 $\neg A \rightarrow [A \rightarrow \neg(A \rightarrow A)]$ is an instance of A7, whereas A10 $[B \rightarrow \neg(A \rightarrow A)] \rightarrow \neg B$ is immediate by Red and A1.

(2) All axioms and rules of B_{KSf} are provable in B_{KS} plus Df: By Proposition 4.16, since B_{cS} is included in B_{KS} (cf. Proposition 3.31 and the diagram following it).   □

Finally, bearing in mind that B_c and B_{KS} are sublogics of B_K, and that B_{cf} and B_{KSf} are sublogics of B_{Kf}, we rely on preceding results to obtain:

Theorem 4.21 (B_K and B_{Kf} are definitionally equivalent). *The logics B_K and B_{Kf} are definitionally equivalent.*

Proof. It follows from the proofs provided in Theorems 4.19 and 4.20.   □

4.4 RM-SEMANTICS FOR THE BASIC CONSTRUCTIVE f-LOGICS

The relationship between the ¬-clause (v) (respectively, (vi)) and the f-clause (vii) (respectively, (ix)) has been discussed in the Introduction. It follows from this discussion that it suffices to change the ¬-clause in the models characterizing a given basic constructive ¬-logic for the appropriate f-clause in order to obtain an RM-semantics for the definitional equivalent f-logic. Thus, we have the following RM-semantics for each one of the four f-logics.

Definition 4.22 (B_{cSf}-models). A B_{cSf}-model is an EB_+-model (K, O, S, R, $\vDash$) where clauses (i)–(iv), (vii) and the semantical postulates P1–P4 and Pf hold.

Definition 4.23 (B_{cf}-models). A B_{cf}-model is an EB_+-model (K, O, R, $\vDash$) where clauses (i)–(iv), (ix) and the semantical postulates P1, P2, and P5 hold.

Definition 4.24 (B_{KSf}-models). A B_{KSf}-model is an EB_{K+}-model (K, S, R, $\vDash$) where clauses (i)–(iv), (vii) and the semantical postulates P1–P3, P6, and Pf hold.

Definition 4.25 (B_{Kf}-models). A B_{Kf}-model is an EB_{K+}-model (K, R, $\vDash$) where clauses (i)–(iv), (ix) and postulates P1, P2, and P7 hold.

Notice that B_{cSf}-models (respectively, B_{KSf}-models) can be defined by restricting $B_{+,f}$-models (respectively, $B_{K+,f}$-models). B_{cf}-models and

B_{Kf}-models, in which $S = K$, are related to some of the models discussed in §6.3.

Given the theorems on definitional equivalence, it is easy to prove that each f-logic is characterized by its respective RM-semantics. As a way of an example, let us show that it is so in the case of the logic B_{cSf} (the proof of the rest of the cases is left to the reader).

We prove that all theorems of B_{cSf} are EB_{cSf}-valid (cf. Definitions 1.5, 1.6, and 1.36).

Proposition 4.26 (All theorems of B_{cSf} are EB_{cSf}-valid). *For each $A \in \mathcal{F}$, if $\vdash_{B_{cSf}} A$, then A is EB_{cSf}-valid.*

Proof. The proof concerning the axioms and rules of B_+ is given in Proposition 1.13. Then, we prove that $A7_f$, Red_f, and $VeqDn_f$ preserve EB_{cSf}-validity.

(a) $A7_f$, $A \Rightarrow f \rightarrow (A \rightarrow f)$, *preserves EB_{cSf}-validity*: Suppose that A is EB_{cSf}-valid and that there is $a \in K$ in some EB_{cSf}-model M such that (1) $a \vDash f$ but (2) $a \nvDash A \rightarrow f$. By 2 and iv, there are $b, c \in K$ in M such that (3) $Rabc$, (4) $b \vDash A$ and (5) $c \nvDash f$. By 5 and vii, we have (6) $c \in S$; by 1 and vii, (7) $a \notin S$. But applying P3, $(Rabc \,\&\, c \in S) \Rightarrow a \in S$, to 3 and 6 we get (8) $a \in S$, contradicting 7.

(b) Red_f, $A \Rightarrow (A \rightarrow f) \rightarrow f$, *preserves EB_{cSf}-validity*: Suppose (1) $\vDash A$ and that there is $a \in K$ in some EB_{cSf}-model M such that (2) $a \vDash A \rightarrow f$ and (3) $a \nvDash f$. By 3 and vii, we have (4) $a \in S$; by P1 and d1, (5) $Rzaa$ for some $z \in O$ in M. Applying now P4 $(Rabc \,\&\, c \in S \Rightarrow \exists x \in O\ \exists y \in S\ Rbxy)$ to 4 and 5, we get (6) $Raxy$ for some $x \in O$ and $y \in S$ in M. By 1, we have (7) $x \vDash A$ since $x \in O$. Finally, by 2, 6, 7, and iv, we get (8) $y \vDash f$ whence, by vii, (9) $y \notin S$ follows, contradicting 6.

(c) $VeqDn_f$, $A \Rightarrow [(A \rightarrow f) \rightarrow f] \rightarrow [B \rightarrow [(A \rightarrow f) \rightarrow f]]$, *preserves EB_{cSf}-validity*: Suppose (1) $\vDash A$ and that there is $B \in \mathcal{F}$ and $a \in K$ in some EB_{cSf}-model M such that (2) $a \vDash (A \rightarrow f) \rightarrow f$ but (3) $a \nvDash B \rightarrow [(A \rightarrow f) \rightarrow f]$. Then, there are $b, c \in K$ in M such that (4) $Rabc$, (5) $b \vDash B$ and (6) $c \nvDash (A \rightarrow f) \rightarrow f$. By 6 and iv there are $d, e \in K$ in M such that (7) $Rcde$, (8) $d \vDash A \rightarrow f$ and (9) $e \nvDash f$. By 9 and vii, we have (10) $e \in S$. Then, by 7, 10, and P4, there are $x \in O$ and $y \in S$ in M such that (11) $Rdxy$. By 1, we have (12) $x \vDash A$, as $x \in O$. Finally, by 8, 11, and 12, we have (13) $y \vDash f$, a contradiction since $y \in S$.   □

Corollary 4.27 (Soundness of B_{cSf}). *For any $A \in \mathcal{F}$, if $\vdash_{B_{cSf}} A$, then $\vDash_{B_{cSf}} A$.*

Proof. Immediate by Proposition 4.26.   □

The proof of the completeness of B_{cSf} easily follows from that of B_{cS}, given the definitional equivalence between both logics. Actually, we only need to prove that clause (vii) holds canonically and that B_{cSf} is a w-consistent logic.

Proposition 4.28 (*a* is w-consistent iff $f \notin a$). *Let L be the logic resulting from deleting $VeqDn_f$ from B_{cSf}. For any EL-theory a, a is w-consistent iff $f \notin a$ (EL is any strengthening of L; cf. Definition 1.36).*

Proof. (a) ($\Rightarrow$) Suppose a is w-consistent. If $f \in a$, then $\neg(A \to A) \in a$ follows by A7$_f$ and D$\neg$, contradicting the w-consistency of a. (b) ($\Leftarrow$) Suppose $f \notin a$. By Red$_f$, we have $A \to f \notin a$, i.e., $\neg A \notin a$, A being a theorem of EL. Thus, a is w-consistent. □

Notice that, given Propositions 4.5 and 4.28, an EL-theory (in the sense of Proposition 4.28) is w-consistent iff it is u-consistent, as it is the case with EB_{cS}-theories (cf. Proposition 2.10).

Proposition 4.29 (B_{cSf} is w-consistent). *The logic B_{cSf} is w-consistent.*

Proof. By Proposition 4.2, f is not B_{cSf}-valid. Then, it follows by the soundness theorem (Corollary 4.27) that f is not derivable in B_{cSf}. Thus, B_{cSf} is w-consistent by Proposition 4.28. □

Definition 4.30 (The canonical B_{cSf}-model). The canonical B_{cSf}-model is a B_+-model of type c, i.e., it is the structure $(K^{NP}, O^P, S^{NP}, R^{NP}, \vDash^{NP})$.

Lemma 4.31 (The canonical model is in fact a model). *The canonical B_{cSf}-model is in fact a model.*

Proof. The proof easily follows from Lemma 2.27 (the canonical B_{cS}-model is a B_{cS}-model) since B_{cS} is included in B_{cSf} (cf. Theorem 4.18 and Definitions D$\neg$ and Df) and consequently all lemmas and propositions proved in Sections 2.3 and 2.4 for EB_{cS}-theories are of course applicable to $B_{cS,f}$-theories. In particular, we have:

(1) The set $O^P \cap S^{NP}$ is not empty: Immediate by Lemma 2.13 since by Proposition 4.29 B_{cSf} is a (regular) and w-consistent theory.

(2) Postulates P1–P4 hold in the canonical B_{cSf}-model: They are proved to hold similarly as in Lemma 2.27

(3) Clauses (i)–(iv) and (vii) are satisfied by the canonical B_{cSf}-model: Clauses (i)–(iv) are proved similarly as in Lemma 2.27. Then, clause (vii) is immediate by Proposition 4.28 proved above. □

Then, completeness is proved in the customary way.

Theorem 4.32 (Completeness of B_{cSf}). *For any $A \in \mathcal{F}$, if $\vDash_{B_{cSf}} A$, then* $\vdash_{B_{cSf}} A$.

Proof. By Lemmas 2.13 and 4.31, Proposition 4.29, and Definitions 4.22 and 4.30. $\square$

The chapter is ended with the theorems noted below establishing that B_{cf}, B_{KSf}, and B_{Kf} are determined by their respective RM-semantics. As pointed out above, the proof of these theorems (left to reader) is similar to those of Corollary 4.27 and Theorem 4.32.

Theorem 4.33 (Soundness and completeness of B_{cf}). *For any $A \in \mathcal{F}$,* $\vDash_{B_{cf}} A$ *iff* $\vdash_{B_{cf}} A$.

Theorem 4.34 (Soundness and completeness of B_{KSf}). *For any $A \in \mathcal{F}$,* $\vDash_{B_{KSf}} A$ *iff* $\vdash_{B_{KSf}} A$.

Theorem 4.35 (Soundness and completeness of B_{Kf}). *For any $A \in \mathcal{F}$,* $\vDash_{B_{Kf}} A$ *iff* $\vdash_{B_{Kf}} A$.

CHAPTER 5

The basic constructive logics $\mathcal{R}B_c$ and $\mathcal{R}B_{c2}$

It has been discussed above that there is not a constructive relevance logic in the RM-semantics if the set of all points does not contain both a subset of designated points and a subset of consistent points. In this chapter, we introduce the logics $\mathcal{R}B_c$ and $\mathcal{R}B_{c2}$, which are two different versions of the basic constructive relevance logic in RM_1-semantics with a set $S \subseteq K$ (RM-semantics with a set of designated points and a set of consistent points). The logic $\mathcal{R}B_c$ is built upon the positive language of B_+ expanded with the unary connective $\neg$; the logic $\mathcal{R}B_{c2}$, upon the language of $\mathcal{R}B_c$ expanded with the propositional truth constant t. A third logic, the logic $\mathcal{R}B_{+,t,f}$, built upon the positive language of B_+ expanded with the two propositional constants, t and f, is also introduced. $\mathcal{R}B_{c2}$ and $\mathcal{R}B_{+,t,f}$ are definitionally equivalent. We recall that "constructive" is understood as "endowed with an intuitionistic-type negation"; and that L is a relevance logic (in the minimal sense) if it has the "variable-sharing property".

5.1 THE LOGIC $\mathcal{R}B_c$

The logic $\mathcal{R}B_c$ is built upon the positive language of B_+ expanded with the unary connective $\neg$ (negation). The logic $\mathcal{R}B_c$ is axiomatized by adding to B_+ the axioms of constructive strong contraposition.

Definition 5.1 (The logic $\mathcal{R}B_c$). The logic $\mathcal{R}B_c$ is axiomatized by adding the following axioms to B_+ (cf. Definition 1.10):

$$A11.\ (A \to \neg B) \to (B \to \neg A)$$
$$A12.\ B \to [(A \to \neg B) \to \neg A]$$

We note that $\mathcal{R}B_c$ is well-axiomatized w.r.t. B_+ and that it is a relevance logic (cf. §5.4).

Some theorems of $\mathcal{R}B_c$ are the following (a proof is sketched to the right of each one of them):

$$T1.\ A \to \neg\neg A \qquad\qquad\qquad \text{A1, A11}$$

Routley-Meyer Ternary Relational Semantics for Intuitionistic-Type Negations.
DOI: http://dx.doi.org/10.1016/B978-0-08-100751-8.00007-9
Copyright © 2018 Elsevier Ltd. All rights reserved.

T2. $(A \rightarrow B) \rightarrow (\neg B \rightarrow \neg A)$	A11, T1
T3. $\neg B \rightarrow [(A \rightarrow B) \rightarrow \neg A]$	A12, T1
T4. $\neg A \rightarrow [A \rightarrow \neg(A \rightarrow A)]$	T1, T2, T3
T5. $[A \rightarrow \neg(B \rightarrow B)] \rightarrow \neg A$	A1, A12
T6. $A \rightarrow [(A \rightarrow \neg B) \rightarrow \neg B]$	A11, A12
T7. $\neg(A \vee B) \leftrightarrow (\neg A \wedge \neg B)$	B$_+$, A11, T2
T8. $(\neg A \vee \neg B) \rightarrow \neg(A \wedge B)$	B$_+$, T2

Notice the following points: T2 and T3 are the axioms of constructive weak contraposition; T4 has been labelled A7′ (cf. Lemma 2.21) or A9 (cf. Definition 3.22) above; the rule Red is immediate by A12; and, finally, A11 and A12 are required for proving T4.

We now move to the definition of an RM-semantics for $\mathcal{R}B_c$.

Definition 5.2 ($\mathcal{R}B_c$-models). An $\mathcal{R}B_c$-model is an EB$_+$-model (K, O, S, $R \vDash$) where clauses (i)–(v) and the following semantical postulates (in addition to P1 and P2) hold

$$P8. \ (R^2 abcd \ \& \ d \in S) \Rightarrow \exists x \in S \ R^2 acbx$$
$$P9. \ (R^2 abcd \ \& \ d \in S) \Rightarrow \exists x \in S \ R^2 bcax$$

(We recall that in all EB$_+$-models, truth in a model and validity are understood according to Definitions 1.5 and 1.6.)

We next prove that all theorems of $\mathcal{R}B_c$ are E$\mathcal{R}B_c$-valid. (The notions of an E$\mathcal{R}B_c$-model and E$\mathcal{R}B_c$-validity are defined according to Definition 1.36.)

Proposition 5.3 (All theorems of $\mathcal{R}B_c$ are E$\mathcal{R}B_c$-valid). *For each $A \in \mathcal{F}$, if $\vdash_{\mathcal{R}B_c} A$, then A is E$\mathcal{R}B_c$-valid.*

Proof. Given that an E$\mathcal{R}B_c$-model is an EB$_+$-model, that the rules of B$_+$ preserve E$\mathcal{R}B_c$-validity, and that the E$\mathcal{R}B_c$-validity of A1–A6 has been proved in Proposition 1.13, it just remains to prove that A11 and A12 are E$\mathcal{R}B_c$-valid (we use Lemma 1.9). We prove the E$\mathcal{R}B_c$-validity of A11, that of A12 is proved similarly.

(a) A11, $(A \rightarrow \neg B) \rightarrow (B \rightarrow \neg A)$, *is E$\mathcal{R}B_c$-valid*: Suppose that there are $A, B \in \mathcal{F}$ and $a \in K$ in some E$\mathcal{R}B_c$-model M such that (1) $a \vDash A \rightarrow \neg B$ but (2) $a \nvDash B \rightarrow \neg A$. By 2 and iv there are $b, c \in K$ in M such that (3) *Rabc*, (4) $b \vDash B$, and (5) $c \nvDash \neg A$. By 5 and v, there are $d, e \in K$ in M such that

(6) $Rcde$, (7) $d \vDash A$, and (8) $e \in S$. By 3, 6, and d2, (9) $R^2 abde$ follows, whence by P8 we have (10) $R^2 adbx$ for some $x \in S$. Then, by applying d2 to 10 we get (11) $Rady$ and (12) $Rybx$ for some $y \in K$. Now, by 1, 7, 11, and iv we obtain (13) $y \vDash \neg B$ whence by 12 we have (14) $b \nvDash B$ as $x \in S$ (by 10). But 4 and 14 contradict each other, Consequently, A11 is $E\mathcal{RB}_c$-valid. $\square$

As it is customary, we have soundness of $\mathcal{RB}_c$ as a corollary of the proposition just proved.

Corollary 5.4 (Soundness of $\mathcal{RB}_c$). *For each $A \in \mathcal{F}$, if $\vdash_{\mathcal{RB}_c} A$, then $\vDash_{\mathcal{RB}_c} A$.*

Proof. Immediate by Proposition 5.3. $\square$

Next, we move to the completeness proof.

Definition 5.5 (The canonical $\mathcal{RB}_c$-*model*). *The canonical $\mathcal{RB}_c$-model is a canonical EB_+-model of type b. That is, it is the structure $(K^P, O^P, S^P, R^P, \vDash^P)$.* (The items of this structure are now referred to $\mathcal{RB}_c$-theories —cf. Definition 1.26.)

The following lemma is needed in the proof that the postulates hold canonically. We prove it for any $E\mathcal{RB}_c$-*logic L (cf. Definition 1.36)*.

Lemma 5.6 (The pre-canonical P8 and P9). *Let L be an $E\mathcal{RB}_c$-logic and $a, b, c \in K^T$, $d \in S^T$ such that $R^{T2} abcd$. Then there are $x, y \in S^P$ such that (a) $R^{T2} acbx$ and (b) $R^{T2} bcay$.*

Proof. We prove case (a) (the proof of case (b) is similar). Given the hypothesis of Lemma 5.6, we have, by d2, (1) $R^T abz$ and $R^T zcd$ for some $z \in K^T$.

Case a: Consider now the sets $y = \{B \mid \exists A[A \to B \in a \ \& \ A \in c]\}$, $u = \{B \mid \exists A[A \to B \in y \ \& \ A \in b]\}$. It is easy to prove that y and u are L-theories such that (2) $R^T acy$ and (3) $R^T ybu$ (cf. Lemma 1.24). Suppose $u \notin S^T$, that is, suppose (4) $\neg A \in u$, A being a theorem of L. By definitions of y and u, we have for some $B, C \in \mathcal{F}$, (5) $C \to (B \to \neg A) \in a$, (6) $B \in b$, and (7) $C \in c$. By A12, we have (8) $\vdash_L (B \to \neg A) \to \neg B$, whence by Pref, we get (9) $\vdash_L [C \to (B \to \neg A)] \to (C \to \neg B)$. So, (10) $C \to \neg B \in a$ follows, whence, by A11, we have (11) $B \to \neg C \in a$. Therefore, (12) $\neg C \in z$ is derivable by 1, 6, and 11. But then we have (13) $C \to \neg(C \to C) \in z$ by T4, whence by 1 and 7, we get (14) $\neg(C \to C) \in d$, contradicting the w-consistency of d. Consequently, u is w-consistent. Then, by Lemma 2.13, u is extended to some $x \in S^P$ such that (15) $u \subseteq x$ and $R^T ybx$. Finally, (16) $R^{T2} acbx$ follows by applying d2 to 2 and 15. $\square$

We can now prove that the canonical $\mathcal{R}B_c$-model is an $\mathcal{R}B_c$-model and the completeness of $\mathcal{R}B_c$.

Lemma 5.7 (The canonical model is a model). *The canonical $\mathcal{R}B_c$-model is in fact an $\mathcal{R}B_c$-model.*

Proof. (1) *The set $O^P \cap S^P$ is not empty*: This follows by using Lemma 2.13 for extending the logic $\mathcal{R}B_c$ to a *w-consistent, regular and prime theory.*

(2) *Postulates P1, P2, P8, and P9 hold in the canonical $\mathcal{R}B_c$-model*: P1 and P2 follow by Corollary 1.23, and P8 and P9 are immediate by Lemma 5.6.

(3) *Clauses (i)–(v) hold in the canonical $\mathcal{R}B_c$-model*: Clauses (i)–(iv) hold by Lemma 1.25, and clause (v) is immediate by Lemmas 2.21 and 2.22.   □

Theorem 5.8 (Completeness of $\mathcal{R}B_c$). *For each $A \in \mathcal{F}$, if $\vDash_{\mathcal{R}B_c} A$, then $\vdash_{\mathcal{R}B_c} A$.*

Proof. By Lemmas 1.17 and 5.7, and Definitions 5.2 and 5.5 (cf. Remark 1.31).   □

5.2 THE LOGIC $\mathcal{R}B_{c2}$

The logic $\mathcal{R}B_{c2}$ is an alternative candidate to $\mathcal{R}B_c$ for being the basic constructive relevance logic in RM_1-semantics with a set $S \subseteq K$ (RM-semantics with a set of designated points and a set of consistent points). The logic $\mathcal{R}B_{c2}$ is built upon the positive language of B_+ expanded with the propositional truth constant t and the unary connective $\neg$ (negation).

Definition 5.9 (The logic $\mathcal{R}B_{c2}$). The logic $\mathcal{R}B_{c2}$ is axiomatized by adding to B_+ (cf. Definition 1.10) the following axioms, rule and definition:

$$A13.\ t$$
$$A14.\ (A \to \neg t) \to \neg A$$
$$A15.\ \neg A \to (A \to \neg t)$$
$$\text{Necessitation (Nec).}\ A \Rightarrow t \to A$$
$$Df.\ f =_{df} \neg t$$

The constant t can intuitively be interpreted as the conjunction of all truths (all theorems). Consequently, the constant f is here intuitively interpreted as the disjunction of all falsehoods in the following strong sense of the term: a falsehood is a formula of the form $\neg A$ where A is a theorem (cf. [2] and [37] on these Ackermann constants, where, we remark, they are used in the context of De Morgan negations; also, notice that f is defined differently in the preceding chapter).

We note that $\mathcal{RB}_{c2}$ is well axiomatized w.r.t. B_+ and that it is a relevance logic (cf. §5.4).

The following are theorems and rules of $\mathcal{RB}_{c2}$. A proof is sketched to the right of each one of them. (Notice that T1–T5 are theorems of $B_{+,f}$.)

T1. $A \to B \Rightarrow \neg B \to \neg A$	Suf, D¬
T2. $\neg B \Rightarrow (A \to B) \to \neg A$	Pref, D¬
T3. $\neg(A \lor B) \leftrightarrow (\neg A \land \neg B)$	t6, D¬
T4. $(\neg A \lor \neg B) \to \neg(A \land B)$	t3, D¬
T5. $\neg f$	A1, D¬
T6. $A \Rightarrow \neg A \to \neg t$	Nec, T1
T7. $A \Rightarrow (B \to \neg A) \to \neg B$	A14, T6, Pref
T8. $A \Rightarrow \neg\neg A$	A1, T7

Next, $\mathcal{RB}_{c2}$-models are defined.

Definition 5.10 ($\mathcal{RB}_{c2}$-models). An $\mathcal{RB}_{c2}$-model is an EB_+-model $(K, O, S, R \vDash)$ where clauses (i)–(v) and (viii) and the following semantical postulates (in addition to P1 and P2) hold

P10. $(Rabc \ \& \ b \in O \ \& \ c \in S) \Rightarrow a \in S$

P11. $(Rabc \ \& \ c \in S) \Rightarrow \exists x \in O \ \exists y \in S \ Rcxy$

We recall that the postulate

$$Pt. \ (a \le b \ \& \ a \in O) \Rightarrow b \in O$$

has to be added (cf. Definition 1.4).

In the following proposition it is proved that all theorems of $\mathcal{RB}_{c2}$ are $E\mathcal{RB}_{c2}$-valid (cf. Definition 1.36).

Proposition 5.11 (All theorems of $\mathcal{RB}_{c2}$ are $E\mathcal{RB}_{c2}$-valid). *For each $A \in \mathcal{F}$, if $\vdash_{\mathcal{RB}_{c2}} A$, then A is $\mathcal{RB}_{c2}$-valid.*

Proof. The axioms and rules of B_+ are proved in Proposition 1.13.

(a) *A13, t, is $E\mathcal{R}B_{c2}$-valid*: Let M be any $E\mathcal{R}B_{c2}$-model and $a \in O$. Then $a \vDash t$ by clause (viii).

(b) *A14, $(A \rightarrow \neg t) \rightarrow \neg A$, is $E\mathcal{R}B_{c2}$-valid*: Suppose that there is $a \in K$ in some $E\mathcal{R}B_{c2}$-model M such that (1) $a \vDash A \rightarrow \neg t$ but (2) $a \nvDash \neg A$. By 2 and v, there are $b, c \in K$ in M such that (3) $Rabc$, (4) $c \in S$, and (5) $b \vDash A$. By 1, 3, 5, and iv, we have (6) $c \vDash \neg t$. By 3, 4, and P11, we have (7) $x \in O$ and (8) $y \in S$ in M such that (9) $Rcxy$. Finally, by 6, 8, 9, and v, we get (10) $x \nvDash t$. But by 7 and viii, (11) $x \vDash t$ follows, a contradiction.

(c) *A15, $\neg A \rightarrow (A \rightarrow \neg t)$, is $E\mathcal{R}B_{c2}$-valid*: Suppose that there is $a \in K$ in some $E\mathcal{R}B_{c2}$-model M such that (1) $a \vDash \neg A$ but (2) $a \nvDash A \rightarrow \neg t$. By 2 and iv there are $b, c \in K$ in M such that (3) $Rabc$, (4) $b \vDash A$, and (5) $c \nvDash \neg t$. By 5 and v, there are $d, e \in K$ such that (6) $Rcde$, (7) $e \in S$, and (8) $d \vDash t$. By 8 and viii, we have (9) $d \in O$. By 6, 7, 9, and P10, we have (10) $c \in S$. Finally, by 1, 3, 10, and v, (11) $b \nvDash A$ follows, contradicting 4.

(d) *Nec, $A \Rightarrow t \rightarrow A$, preserves $E\mathcal{R}B_{c2}$-validity*: Suppose (1) $\vDash A$, i.e., that A is $E\mathcal{R}B_{c2}$-valid, but that there is $a \in K$ in some $E\mathcal{R}B_{c2}$-model such that (2) $a \vDash t$ and (3) $a \nvDash A$. By 2 and viii, we have (4) $a \in O$. Then, we get (5) $a \vDash A$ by 1 and 4 (cf. Definition 1.6), a contradiction. $\square$

Corollary 5.12 (Soundness of $\mathcal{R}B_{c2}$). *For each $A \in \mathcal{F}$, if $\vdash_{\mathcal{R}B_{c2}} A$, then $\vDash_{\mathcal{R}B_{c2}} A$.*

Proof. Immediate by Proposition 5.11. $\square$

We remark the following proposition.

Proposition 5.13 (Negation of theorems are not $E\mathcal{R}B_{c2}$-valid). *Let A be a theorem of $\mathcal{R}B_{c2}$. Then $\neg A$ is false in every $a \in O \cap S$ in each $E\mathcal{R}B_{c2}$-model.*

Proof. Suppose that A is a theorem of $\mathcal{R}B_{c2}$ and that there is $a \in O \cap S$ in some $E\mathcal{R}B_{c2}$-model M such that (1) $a \vDash \neg A$. (Recall that $O \cap S$ is not empty; cf. Definition 1.4.) By P1 and d1, there is (2) $x \in O$ such that $Rxaa$. By P11, there are (3) $y \in O$ and (4) $z \in S$ such that (5) $Rayz$. By 1, 4, 5, and v, (6) $y \nvDash A$ follows. But as A is a theorem, A is true in every $b \in O$ by the soundness theorem (Corollary 5.12), and consequently, (7) $y \vDash A$ follows by 3, contradicting 6. $\square$

In the following pages we proceed to the completeness theorem. We begin by proving some preliminary propositions in which (as well as in the rest of this section) it has to be understood that by an L-logic we shall refer to an $E\mathcal{RB}_{c2}$-logic, not simply to an EB_+-logic. Consequently, the terms K^T, O^T, S^T etc. have to be understood in the same sense (cf. Definition 1.16).

Proposition 5.14 ($a \in O^T$ iff $t \in a$). *Let $a \in K^T$. Then, $a \in O^T$ iff $t \in a$.*

Proof. (a) ($\Rightarrow$) If a is regular, then $t \in a$ by A13. (b) ($\Leftarrow$) Suppose $t \in a$ and let A be a theorem. By Nec, $t \to A$ is also a theorem and so we have $A \in a$, that is, any theorem A belongs to a if t belongs to it, as it was required. $\square$

Proposition 5.15 ($a \in S^T$ iff $f \notin a$). *Let $a \in K^T$. Then, $a \in S^T$ iff $f \notin a$.*

Proof. We use Df. (a) ($\Rightarrow$) Suppose $a \in S^T$. If $\neg t \in a$, then a is w–inconsistent by A13. (b) ($\Leftarrow$) Suppose $\neg t \notin a$ but $\neg A \in a$, A being a theorem of L. By T6, $\neg A \to \neg t$ is a theorem. Then, we have $\neg t \in a$, contradicting the hypothesis. $\square$

Lemma 5.16 (The pre-canonical P10). *Let $a \in K^T$, $b \in O^T$ and $c \in S^T$ such that $R^T abc$. Then, $a \in S^T$.*

Proof. Suppose (1) $a \in K^T$, (2) $b \in O^T$, and (3) $c \in S^T$ such that (4) $R^T abc$. Further, suppose (5) $a \notin S^T$, i.e., $\neg t \in a$ (cf. Df and Proposition 5.15). By A15, we have (6) $\neg t \to (t \to \neg t)$ and by 4, 5, and 6, (7) $t \to \neg t \in a$. Now, (8) $t \in b$ follows by 2 and A13. Then, we get (9) $\neg t \in c$ by 4, 7, and 8, contradicting the consistency of c. $\square$

Lemma 5.17 (The pre-canonical P11). *Let $a, b \in K^T$ and $c \in S^T$ such that $R^T abc$. Then, there are $x \in O^P$ and $y \in S^P$ such that $R^T cxy$.*

Proof. Let $a, b \in K^T$ and $c \in S^T$ such that $R^T abc$. Consider the sets $z = \{A \mid \vdash_L A\}$ and $u = \{B \mid \exists A[A \to B \in c\ \&\ A \in x]\}$. It is not difficult to prove that z and u are L-theories such that $R^T czu$. Clearly z is regular and next u is proved w-consistent. Suppose it is not. Then, we have $f \in u$ (Proposition 5.15), whence we get $A \to f \in c$, A being a theorem of L; by A14 and Df, we have $\neg A \in c$, contradicting the w-consistency of c. Finally, z and u are extended to the required prime L-theories x and y similarly as in Lemma 2.18. $\square$

Lemmas 5.18 and 5.19 below are proved by using A14, $(A \to \neg t) \to \neg A$, and A15, $\neg A \to (A \to \neg t)$, respectively, similarly as A7′ $\neg A \to [A \to \neg (A \to A)]$ and Red $A \Rightarrow (B \to \neg A) \to \neg B$ are used in the proof of Lemma 2.21 and Lemma 2.22.

Lemma 5.18 (The canonical negation clause I). *For any $A \in \mathcal{F}$ and $a \in K^T$, if $\neg A \in a$, then for any $b, c \in K^T$, $(R^T abc \ \& \ c \in S^T) \Rightarrow A \notin b$.*

Proof. Suppose $\neg A \in a$ and $R^T abc$ for $a, b \in K^T$ and $c \in S^T$. Moreover, for reductio, suppose $A \in b$. By A15, we get $A \rightarrow \neg t \in a$ and, finally, we have $\neg t \in c$, contradicting the w-consistency of c. $\square$

Lemma 5.19 (The canonical negation clause II). *For any $A \in \mathcal{F}$ and $a \in K^T$, if $\neg A \in a$, there are $x \in K^P$ and $y \in S^P$ such that $R^P axy$ and $A \in x$.*

Proof. Suppose $\neg A \notin a$ for $A \in \mathcal{F}$ and $a \in K^T$. The sets $z = \{B \mid \vdash_L A \rightarrow B\}$ and $u = \{C \mid \exists D[D \rightarrow C \in a \ \& \ D \in z]\}$ are L-theories such that $R^T azu$ and $A \in z$ (by A1). It remains to prove that u is w-consistent. Suppose that u is not w-consistent. That is, suppose $\neg t \in u$ (cf. Proposition 5.15 and Df). Then, we have $B \rightarrow \neg t \in a$, $\vdash_L A \rightarrow B$ for some $B \in \mathcal{F}$. By the rule Suf $\vdash_L (B \rightarrow \neg t) \rightarrow (A \rightarrow \neg t)$ follows. So, we get $A \rightarrow \neg t \in a$, whence, by Red, we have $\neg A \in a$, contradicting the hypothesis. Then, u and z are extended in this order to the required prime L-theories y and x by Lemmas 2.13 and 1.19 (cf. the proof of Lemma 2.22). $\square$

Proposition 5.20 ($\mathcal{RB}_{c2}$ is w-consistent). *The logic $\mathcal{RB}_{c2}$ is w-consistent.*

Proof. Let A be a theorem of $\mathcal{RB}_{c2}$. By Proposition 5.13, $\neg A$ is not $\mathcal{RB}_{c2}$-valid. Then $\neg A$ is not derivable in $\mathcal{RB}_{c2}$ by the soundness theorem (Corollary 5.12). Thus, $\mathcal{RB}_{c2}$ is w-consistent. $\square$

Next, we prove the completeness theorem once the canonical model has been defined.

Definition 5.21 (The canonical $\mathcal{RB}_{c2}$-model). The canonical $\mathcal{RB}_{c2}$-model is a canonical EB$_+$-model of type b. That is, it is the structure $(K^P, O^P, S^P, R^P, \vDash^P)$ where the items K^P, O^P, S^P, R^P, and $\vDash^P$ are referred to $\mathcal{RB}_{c2}$-theories (cf. Definition 1.26).

Lemma 5.22 (The canonical model is a model). *The canonical $\mathcal{RB}_{c2}$-model is in fact a $\mathcal{RB}_{c2}$-model.*

Proof. (1) *The set $O^P \cap S^P$ is not empty*: As shown in Proposition 5.20, the logic $\mathcal{RB}_{c2}$ is w-consistent. Then, by Lemma 2.13, the logic $\mathcal{RB}_{c2}$ is extended to a w-consistent, regular and prime $\mathcal{RB}_{c2}$-theory.

(2) *Postulates P1, P2, P10, P11, and Pt hold in the canonical $\mathcal{RB}_{c2}$-model*: P1, P2, and Pt are immediate by Corollary 1.23, and P10 and P11 are immediate by Lemmas 5.16 and 5.17, respectively.

(3) *Clauses (i)–(iv), (v), and (viii) hold in the canonical $\mathcal{RB}_{c2}$-model:* Clauses (i)–(iv) are proved by Lemma 1.25; clause (viii) by Proposition 5.14. Then, clause (v) is immediate by Lemmas 5.18 and 5.19.   □

Theorem 5.23 (Completeness of $\mathcal{RB}_{c2}$). *For any $A \in \mathcal{F}$, if $\vDash_{\mathcal{RB}_{c2}} A$, then $\vdash_{\mathcal{RB}_{c2}} A$.*

Proof. By Definitions 5.10 and 5.21, and Lemmas 2.13 and 5.22.   □

5.3 THE LOGIC $\mathcal{RB}_{+,t,f}$

The logic $\mathcal{RB}_{+,t,f}$ is built upon the positive language of $\mathcal{RB}_+$ expanded with the propositional truth constant t and the propositional falsity constant f. Thus, it is an expansion of the logic $B_{+,f}$. The logic $\mathcal{RB}_{+,t,f}$ is defined as follows.

Definition 5.24 (The logic $\mathcal{RB}_{+,t,f}$). The logic $\mathcal{RB}_{+,t,f}$ is axiomatized by adding the following axioms, rule and definition to B_+ (cf. Definition 1.10):

$$A13.\ t$$
$$A16.\ (t \rightarrow f) \rightarrow f$$
$$A17.\ f \rightarrow (t \rightarrow f)$$
$$\text{Necessitation (Nec).}\ A \Rightarrow t \rightarrow A$$
$$D\neg.\ \neg A =_{\mathrm{df}} A \rightarrow f$$

The logic $\mathcal{RB}_{+,t,f}$ is definitionally equivalent to $\mathcal{RB}_{c2}$, as it is shown below.

Proposition 5.25 (Provability of $D\neg$ and Df). *(a) $D\neg$ is provable in $\mathcal{RB}_{c2} \cup \{Df\}$; (b) Df is provable in $\mathcal{RB}_{+,t,f} \cup \{D\neg\}$.*

Proof. (a) By A14, A15, and Df; (b) by A16, A17, and $D\neg$.   □

Proposition 5.26 ($\mathcal{RB}_{c2} \subseteq \mathcal{RB}_{+,t,f} \cup \{D\neg\}$; $\mathcal{RB}_{+,t,f} \subseteq \mathcal{RB}_{c2} \cup \{Df\}$). *(a) All axioms and rules of $\mathcal{RB}_{c2}$ are provable in $\mathcal{RB}_{+,t,f} \cup \{D\neg\}$; (b) all axioms and rules of $\mathcal{RB}_{+,t,f}$ are provable in $\mathcal{RB}_{c2} \cup \{Df\}$.*

Proof. (a) By A16, A17, and $D\neg$; (b) by A14, A15, and Df.   □

Then, we have:

Theorem 5.27 ($\mathcal{R}B_{c2}$ and $\mathcal{R}B_{+,t,f}$ are definitionally equivalent). *The logics $\mathcal{R}B_{c2}$ and $\mathcal{R}B_{+,t,f}$ are definitionally equivalent.*

Proof. By Propositions 5.25 and 5.26.   □

We end the section by presenting an RM-semantics for $\mathcal{R}B_{+,t,f}$.

Definition 5.28 ($\mathcal{R}B_{+,t,f}$-models). An $\mathcal{R}B_{+,t,f}$-model is an EB_+-model $(K, O, S, R, \vDash)$ where clauses (i)–(iv), (vii), and (viii) and the following semantical postulates hold: P1, P2, P10, P11, Pf, and Pt.

We leave to reader the proof of the following theorem:

Theorem 5.29 (Soundness and completeness of $\mathcal{R}B_{+,t,f}$). *For any $A \in \mathcal{F}$, $\vDash_{\mathcal{R}B_{+,t,f}} A$ iff $\vdash_{\mathcal{R}B_{+,t,f}} A$.*

The proof can be carried out similarly as in the case of $\mathcal{R}B_{c2}$; or else one can lean upon the definitional equivalence between $\mathcal{R}B_{c2}$ and $\mathcal{R}B_{+,t,f}$, as with the f-logics in Chapter 4.

5.4 INDEPENDENCE AND VARIABLE-SHARING PROPERTY IN $\mathcal{R}B_c$ AND $\mathcal{R}B_{c2}$

As it is known, Anderson and Belnap's relevance logic R can be expanded conservatively by adding to the language of R the propositional truth constant t together with the rules tI (if A is a theorem, so is $t \to A$) and tE (if $t \to A$ is a theorem, so is A (cf. [3], §R2)). This expansion Rt is a relevance logic in the sense that it has the variable-sharing property (vsp). Consequently, $\mathcal{R}B_c$ *and* $\mathcal{R}B_{c2}$ *have the vsp* (concerning the vsp in Rt and $\mathcal{R}B_{c2}$, consult [43]).

Next, in Propositions 5.30 and 5.31 below, it is shown that $\mathcal{R}B_c$ and $\mathcal{R}B_{c2}$ are well axiomatized w.r.t. B_+.

Proposition 5.30 (Independence in $\mathcal{R}B_c$). *The logic $\mathcal{R}B_c$ is well-axiomatized w.r.t. B_+. That is, given B_+, A11, and A12 are independent from each other.*

Proof. Consider the following sets of truth tables Set I and Set II, where designated values are starred. Both sets verify the axioms and rules of B_+. In addition, Set I verifies A11 and Set II verifies A12.

Set I (Independence of A12):

→	0	1	2	¬		∧	0	1	2		∨	0	1	2
0	1	1	1	2		0	0	0	0		0	0	1	2
*1	0	1	1	0		*1	0	1	1		*1	1	1	2
*2	0	0	1	0		*2	0	1	2		*2	2	2	2

Falsifies A12 when $A = 0$ and $B = 2$.

Set II (Independence of A11):

→	0	1	2	3	4	¬		∧	0	1	2	3	4
0	4	4	4	4	4	4		0	0	0	0	0	0
1	1	3	4	4	4	3		1	0	1	1	1	1
2	1	2	3	3	4	2		2	0	1	2	2	2
*3	0	1	2	3	4	1		*3	0	1	2	3	3
*4	0	0	0	2	4	0		*4	0	1	2	3	4

∨	0	1	2	3	4
0	0	1	2	3	4
1	1	1	2	3	4
2	2	2	2	3	4
*3	3	3	3	3	4
*4	4	4	4	4	4

Falsifies A11 when $A = 4$ and $B = 1$.   □

Proposition 5.31 (Independence in $\mathcal{RB}_{c2}$). *The logic $\mathcal{RB}_{c2}$ is well-axiomatized w.r.t. B_+. That is, given B_+, A13, A14, A15, and Nec are independent from each other.*

Proof. Consider the following sets of truth-tables, Set I–Set IV, where designated values are starred. The four sets verify the axioms and rules of B_+. In addition, Set I verifies A13, A14, and A15; Set II, A13, A14, and Nec; Set III, A13, A15, and Nec, and, finally, Set IV, A14, A15, and Nec.

Set I (Independence of Nec): It is the same set as Set I in Proposition 2.38. t is assigned the value 2 and Nec is falsified when $A = 1$.

Set II (Independence of A15): It is the same set as Set II in Proposition 2.38. t is assigned the value 1 and A15 is falsified when $A = 0$.

Set III (Independence of A14): It is the same set as Set III in Proposition 2.38. t is assigned the value 1 and A14 is falsified when $A = 0$.

Set IV (Independence of t):

→	0	1	¬		∧	0	1		∨	0	1
0	1	1	1		0	0	0		0	0	1
*1	0	1	1		*1	0	1		*1	1	1

t is assigned the value 0.   □

Let us end the chapter by remarking the following. (1) Although included in relevance logic R, $\mathcal{R}B_c$ and $\mathcal{R}B_{c2}$ are not included in Anderson and Belnap's relevance logics of entailment such as E (*Entailment*) and T (*Ticket Entailment*). (2) In Chapters 6 and 7 the reader can find a wealth of strong constructive relevance logics extending $\mathcal{R}B_c$ and $\mathcal{R}B_{c2}$, several of which are not included in R and consequently cannot be built by expanding the language of R with f and then defining negation as in $\mathcal{R}B_{c2}$. (3) $\mathcal{R}B_c$ and $\mathcal{R}B_{c2}$ are the basic constructive relevance logics in RM_1-semantics in the following sense. (a) $\mathcal{R}B_c$: As shown in Lemmas 2.21 and 2.22, A7' and Red are required in the proof that the negation clause holds canonically. But the postulate P3 (that validates A7') cannot be proved in canonical models for relevance logics, i.e., canonical EB_+-models of type a and b. However, A7' can be proved as shown above, with A11 and A12, the corresponding postulates of which are provable in canonical EB_+-models of type b. (b) $\mathcal{R}B_{c2}$: A14 and A15 (respectively, A13 and Nec) are needed in the proof of the canonical adequacy of the negation clause (respectively, clause (viii)). (4) Strong soundness and completeness theorems (of sorts) can be proved for the logics defined in this chapter (cf. Remark 1.31).

CHAPTER 6

Extensions and expansions of the basic logics

In this chapter, we investigate a wealth of extensions and expansions of the basic logics B_+, B_{K+}, B_{cS}, B_{KS}, B_c, B_K, B_{cSf}, B_{KSf}, B_{cf}, B_{Kf}, $\mathcal{R}B_c$, $\mathcal{R}B_{c2}$ and $\mathcal{R}B_{+,t,f}$.

These extensions and expansions are defined from the set of schemes t1–t71 that are displayed below. This set contains several well-known theses of propositional logic leaning on which it is shown how to model different kinds of intuitionistic-type negations in RM-semantics. Given that an RM-semantics for an EB_+-logic L is provided when L-models together with the notion of L-validity are defined (cf. Definition 1.7), the idea is to give a semantical postulate corresponding to each one of the schemes t1–t71. Then soundness and completeness theorems for extensions and expansions of EB_+-logics with any set of these schemes are immediate (cf. Theorems 6.10 and 6.17). Some particular extensions and expansions of the basic logics are briefly treated in the following chapter. Of course, there are many other theses that we have not considered and that could be interpreted in RM-semantics and then used to build extensions and expansions of EB_+-logics similarly as t1–t71 are going to be interpreted and used. The work on these possible additional theses is left to the readers of the present book.

6.1 EXTENSIONS AND EXPANSIONS WITH POSITIVE AXIOMS

We begin the study of extensions and expansions of the basic logics by investigating those defined by means of a set of theses and an important rule formulated in the positive (i.e., negationless) language of B_+ and B_{K+}.

Consider the following theses and rule:

List of positive axioms:

> t1. $[(A \to B) \land (B \to C)] \to (A \to C)$
>
> t2. $(B \to C) \to [(A \to B) \to (A \to C)]$
>
> t3. $(A \to B) \to [(B \to C) \to (A \to C)]$

Routley-Meyer Ternary Relational Semantics for Intuitionistic-Type Negations.
DOI: http://dx.doi.org/10.1016/B978-0-08-100751-8.00008-0

t4. $[A \wedge (A \rightarrow B)] \rightarrow B$

t5. $[A \rightarrow (A \rightarrow B)] \rightarrow (A \rightarrow B)$

t6. $A \rightarrow [[A \rightarrow (A \rightarrow B)] \rightarrow B]$

t7. $[A \rightarrow (B \rightarrow C)] \rightarrow [(A \rightarrow B) \rightarrow (A \rightarrow C)]$

t8. $(A \rightarrow B) \rightarrow [[A \rightarrow (B \rightarrow C)] \rightarrow (A \rightarrow C)]$

t9. $[A \rightarrow (B \rightarrow C)] \rightarrow [(A \wedge B) \rightarrow C]$

t10. $[[(A \rightarrow A) \wedge (B \rightarrow B)] \rightarrow C] \rightarrow C$

t11. $A \Rightarrow (A \rightarrow B) \rightarrow B$

t12. $A \rightarrow [[A \rightarrow (B \rightarrow C)] \rightarrow (B \rightarrow C)]$

t13. $[A \rightarrow [B \rightarrow (C \rightarrow D)]] \rightarrow [B \rightarrow [A \rightarrow (C \rightarrow D)]]$

t14. $A \rightarrow [(A \rightarrow B) \rightarrow B]$

t15. $[A \rightarrow (B \rightarrow C)] \rightarrow [B \rightarrow (A \rightarrow C)]$

t16. $(A \wedge B) \rightarrow [[A \rightarrow (B \rightarrow C)] \rightarrow C]$

t17. $(A \rightarrow B) \rightarrow [[A \wedge (B \rightarrow C)] \rightarrow C]$

t18. $[A \wedge (B \rightarrow C)] \rightarrow [(A \rightarrow B) \rightarrow C)]$

t19. $B \rightarrow [[A \rightarrow (B \rightarrow C)] \rightarrow (A \rightarrow C)]$

t20. $A \rightarrow (A \rightarrow A)$

t21. $A \rightarrow [B \rightarrow (A \vee B)]$

t22. $(A \rightarrow B) \vee (B \rightarrow A)$

t23. $[A \rightarrow (B \vee C)] \rightarrow [(A \rightarrow B) \vee (A \rightarrow C)]$

t24. $[(A \wedge B) \rightarrow C] \rightarrow [(A \rightarrow C) \vee (B \rightarrow C)]$

t25. $B \rightarrow (A \rightarrow A)$

t26. $(A \rightarrow B) \rightarrow [C \rightarrow (A \rightarrow B)]$

t27. $A \rightarrow (B \rightarrow A)$

t28. $A \rightarrow [B \rightarrow (C \rightarrow A)]$

t29. $(A \vee B) \rightarrow [(A \rightarrow B) \rightarrow B]$

t30. $A \rightarrow [B \rightarrow (A \wedge B)]$

t31. $[(A \wedge B) \rightarrow C] \rightarrow [A \rightarrow (B \rightarrow C)]$

t32. $(A \rightarrow B) \vee [(A \rightarrow B) \rightarrow C]$

t33. $A \vee (A \rightarrow B)$

t34. $A \vee [A \rightarrow [B \vee (B \rightarrow C)]]$

t35. $A \rightarrow [B \vee [(A \rightarrow B) \rightarrow C]]$

t36. $[A \vee (B \rightarrow C)] \vee (A \rightarrow B)$

t37. $(A \rightarrow C) \vee (B \rightarrow A)$

t38. $A \vee [B \rightarrow (A \rightarrow C)]$

t39. $B \rightarrow [(B \rightarrow C) \vee (A \rightarrow B)]$

t40. $B \rightarrow [A \vee (A \rightarrow C)]$

Consider now the following semantical postulates (corresponding to t1–t40) where, unless otherwise stated, quantifiers range over K and (in addition to the definitions d1 and d2 —cf. Definitions 1.4 and 3.3) the following definition holds:

$$R^3 abcde =_{df} \exists x, y(Rabx \ \& \ Rxcy \ \& \ Ryde)$$

List of positive postulates:

pt1. $Rabc \Rightarrow \exists x(Rabx \ \& \ Raxc)$

pt2. $R^2 abcd \Rightarrow \exists x(Rbcx \ \& \ Raxd)$

pt3. $R^2 abcd \Rightarrow \exists x(Racx \ \& \ Rbxd)$

pt4. $Raaa$

pt5. $Rabc \Rightarrow R^2 abbc$

pt6. $Rabc \Rightarrow R^2 baac$

pt7. $R^2 abcd \Rightarrow \exists x, y(Racx \ \& \ Rbcy \ \& \ Rxyd)$

pt8. $R^2 abcd \Rightarrow \exists x, y(Racx \ \& \ Rbcy \ \& \ Ryxd)$

pt9. $Rabc \Rightarrow R^2 abbc$

pt10. $\exists x \in Z \ Raxa$

        [Za iff for all $b, c \in K$, $Rabc \Rightarrow \exists x \in O \ Rxbc$]

pt11. $\exists x \in O \ Raxa$

pt12. $R^2 abcd \Rightarrow R^2 bacd$

pt13. $R^3 abcde \Rightarrow R^3 acbde$

pt14. $Rabc \Rightarrow Rbac$

pt15. $R^2 abcd \Rightarrow R^2 acbd$

pt16. $Rabc \Rightarrow R^2 baac$

pt17. $Rabc \Rightarrow \exists x(Rabx \ \& \ Rbxc)$

pt18. $Rabc \Rightarrow \exists x(Rbax \ \& \ Raxc)$

pt19. $R^2abcd \Rightarrow R^2bcad$

pt20. $Rabc \Rightarrow (a \leq c$ or $b \leq c)$

pt21. $Rabc \Rightarrow (a \leq c$ or $b \leq c)$

pt22. $(Rabc$ & $Rade$ & $a \in O) \Rightarrow (b \leq e$ or $d \leq c)$

pt23. $(Rabc$ & $Rade) \Rightarrow \exists x[(Rabx$ or $Radx)$ & $x \leq c$ & $x \leq e]$

pt24. $(Rabc$ & $Rade) \Rightarrow \exists x[(Raxc$ or $Raxe)$ & $b \leq x$ & $d \leq x]$

pt25. $Rabc \Rightarrow b \leq c$

pt26. $R^2abcd \Rightarrow Racd$

pt27. $Rabc \Rightarrow a \leq c$

pt28. $R^2abcd \Rightarrow a \leq d$

pt29. $Rabc \Rightarrow (Rbac$ & $a \leq c)$

pt30. $Rabc \Rightarrow (a \leq c$ & $b \leq c)$

pt31. $R^2abcd \Rightarrow \exists x(Raxd$ & $b \leq x$ & $c \leq x)$

pt32. $(Rabc$ & $Rade$ & $a \in O) \Rightarrow Rdbc$

pt33. $(Rabc$ & $a \in O) \Rightarrow b \leq a$

pt34. $(Rabc$ & $Rcde$ & $a \in O) \Rightarrow (b \leq a$ or $d \leq c)$

pt35. $Rabc \Rightarrow \exists x(Rbax$ & $x \leq a)$

pt36. $(Rabc$ & $Rade$ & $a \in O) \Rightarrow (d \leq a$ or $b \leq e)$

pt37. $(Rabc$ & $Rade$ & $a \in O) \Rightarrow b \leq e$

pt38. $(Rabc$ & $Rcde$ & $a \in O) \Rightarrow d \leq a$

pt39. $(Rabc$ & $Rade) \Rightarrow (a \leq e$ or $b \leq e)$

pt40. $Rabc \Rightarrow b \leq a$

The set O is not defined in EB_{K+}-models (cf. Definition 3.3). Consequently, all references to this set in pt1–pt40 must be dropped when these postulates are used with EB_{K+}-models. Thus, for example, pt10, pt11, and pt32 are read, respectively, as follows in EB_{K+}-models: $\exists x Raxa$, $\exists x Raxa$, $(Rabc$ & $Rade) \Rightarrow Rdbc$.

We shall define extensions of the basic logics by using t1–t40, and RM-semantics for these extensions by means of the corresponding postulates pt1–pt40. We note that t1–t5, t7–t9, t11, t14, t15, t20, t22, t24–t31, and t33 are among the theses and rules considered in [37] for extending Routley and Meyer's basic logic B.

Firstly, we prove the validity of t1–t40 in any EB_+-model or EB_{K+}-model in which the corresponding semantical postulate holds:

Proposition 6.1 (Validity of t1–t40). *Let $\mathcal{M}$ be a class of EB_+-models or EB_{K+}-models and $M \in \mathcal{M}$. Then, for any $j(1 \le j \le 10; 12 \le j \le 40)$, tj is true in M if ptj holds in M; and t11 preserves truth in M if pt11 holds in M.*

Proof. The proof is similar to that provided in [37], Chapter 4, for the extensions of B referred to above or, indeed, to the soundness proofs provided in the preceding chapters. So, it will suffice to prove some selected items. (In this and the rest of the soundness proofs to follow, we lean upon the Entailment Lemma —cf. Lemmas 1.9 and 3.9— and use the Hereditary Condition Lemma (cf. Lemmas 1.8 and 3.8). Also recall that by iv, v, and vi we refer to clauses (iv), (v), and (vi) in Definitions 1.4 and 3.3.)

(a) *t7, $[A \rightarrow (B \rightarrow C)] \rightarrow [(A \rightarrow B) \rightarrow (A \rightarrow C)]$, is true in M:* Suppose there are $A, B, C \in \mathcal{F}$ and $a \in K$ in M such that (1) $a \vDash A \rightarrow (B \rightarrow C)$ but (2) $a \nvDash (A \rightarrow B) \rightarrow (A \rightarrow C)$. By 2 and iv, there are $b, c \in K$ such that (3) $Rabc$, (4) $b \vDash A \rightarrow B$, and (5) $c \nvDash A \rightarrow C$. By 5 and iv, there are $d, e \in K$ such that (6) $Rcde$, (7) $d \vDash A$, and (8) $e \nvDash C$. By d2, 3, and 6, we have (9) R^2abde; and by pt7, (10) $Radx$, (11) $Rbdy$, and (12) $Rxye$ for $x, y \in K$. Then, the following consequences are derivable by applying clause iv: (13) $x \vDash B \rightarrow C$, by 1, 7, and 10; (14) $y \vDash B$, by 4, 7, and 11; (15) $e \vDash C$, by 12, 13, and 14. But 8 and 15 contradict each other.

(b) *t10, $[[(A \rightarrow A) \wedge (B \rightarrow B)] \rightarrow C] \rightarrow C$, is true in M:* Suppose there are $A, B, C \in \mathcal{F}$ and $a \in K$ in M such that (1) $a \vDash [(A \rightarrow A) \wedge (B \rightarrow B)] \rightarrow C$ but (2) $a \nvDash C$. By pt10, there is some $x \in Z$ such that (3) $Raxa$. Then (4) $x \nvDash A \rightarrow A$ or $x \nvDash B \rightarrow B$ follows by 1, 2, and 3. Suppose (5) $x \nvDash A \rightarrow A$. Then, there are $b, c \in K$ in M such that (6) $Rxbc$, (7) $b \vDash A$, and (8) $c \nvDash A$. By pt10, there is some $y \in O$ such that (9) $Rybc$. By 9 and d1, we have (10) $b \le c$, whence by 7, we conclude (11) $c \vDash A$, a contradiction. The supposition $x \nvDash B \rightarrow B$ also leads to a contradiction, in a similar way.

(c) *t11, If A then $(A \rightarrow B) \rightarrow B$, preserves truth in M:* Suppose there are $A, B \in \mathcal{F}$ and $a \in O$ in M such that (1) A is true in M but (2) $a \nvDash (A \rightarrow B) \rightarrow B$. By 2 and iv, there are $b, c \in K$ such that (3) $Rabc$, (4) $b \vDash A \rightarrow B$, and (5) $c \nvDash B$. By pt11 there is some $x \in O$ in M such that (6) $Rbxb$. Clearly (7) $x \vDash A$ since A is true in M and $x \in O$. Then, we have (8) $b \vDash B$ by 4, 6, 7, and iv. On the other hand, (9) $b \le c$ follows by d1 and 3 as $a \in O$. Thus, we have (10) $c \vDash B$ by 8 and 9. But 5 and 10 contradict each other.

(d) *t13, $[A \rightarrow [B \rightarrow (C \rightarrow D)]] \rightarrow [B \rightarrow [A \rightarrow (C \rightarrow D)]]$, is true in M:* Suppose there are $A, B, C, D \in \mathcal{F}$ and $a \in K$ in M such that (1) $a \vDash A \rightarrow [B \rightarrow (C \rightarrow D)]$ but (2) $a \nvDash B \rightarrow [A \rightarrow (C \rightarrow D)]$. By applying clause (iv), we have: (3) $b \vDash B$ and (4) $c \nvDash A \rightarrow (C \rightarrow D)$ for $b, c \in K$ such that (5) $Rabc$;

next, (6) $d \vDash A$ and (7) $e \nvDash C \to D$ for $d, e \in K$ such that (8) $Rcde$ and, finally, (9) $g \vDash C$ and (10) $h \nvDash D$ for $g, h \in K$ such that (11) $Regh$. By 5, 8, 11, and d3, we get (12) R^3abdgh; and, by pt13, (13) R^3adbgh, whence by d3 we get, for $x, y \in K$, (14) $Radx$, (15) $Rxby$, and (16) $Rygh$. Then, the following consequences are derivable: (17) $x \vDash B \to (C \to D)$, by 1, 6, and 14; (18) $y \vDash C \to D$, by 3, 15, and 17 and, finally, (19) $h \vDash D$, by 9, 16, and 18. But 10 and 19 contradict each other.

(e) $t20$, $A \to (A \to A)$, is true in M: Suppose there are $A \in \mathcal{F}$ and $a \in K$ in M such that (1) $a \vDash A$ but (2) $a \nvDash A \to A$. Then there are $b, c \in K$ in M such that (3) $Rabc$, (4) $b \vDash A$, and (5) $c \nvDash A$. By 3 and pt20, we have (6) $a \leq c$ or (7) $b \leq c$. By 1 and 6 or 2 and 7, we get (8) $c \vDash A$, contradicting 5.

(f) $t24$, $[(A \wedge B) \to C] \to [(A \to C) \vee (B \to C)]$, is true in M: Suppose there are $A, B, C \in \mathcal{F}$ and $a \in K$ in M such that (1) $a \vDash (A \wedge B) \to C$ but (2) $a \nvDash (A \to C) \vee (B \to C)$. Then, we have (3) $a \nvDash A \to C$ and (4) $a \nvDash B \to C$ and so there are $b, c \in K$ such that (5) $Rabc$, (6) $b \vDash A$, and (7) $c \nvDash C$; and there are $d, e \in K$ such that (8) $Rade$, (9) $d \vDash B$, and (10) $e \nvDash C$. By pt24, 5, and 8, there is some $x \in K$ such that (11) $Raxc$ or (12) $Raxe$, (13) $b \leq x$, and (14) $d \leq x$. Now, by 6, 9, 13, and 14 we have (15) $x \vDash A \wedge B$. Next, suppose 11. Then, (16) $c \vDash C$ follows by 1 and 15, contradicting 7. On the other hand, if we suppose 12, then we have (17) $e \vDash C$, by 1 and 15, contradicting 10.

(g) $t25$, $B \to (A \to A)$, is true in M: Suppose there are $A, B \in \mathcal{F}$ and $a \in K$ in M such that (1) $a \vDash B$ but (2) $a \nvDash A \to A$. Then, there are $b, c \in K$ such that (3) $Rabc$, (4) $b \vDash A$, and (5) $c \nvDash A$. By 3 and pt25, (6) $b \leq c$ follows, whence by 4, we have (7) $c \vDash A$, contradicting 5.

(h) $t31$, $[(A \wedge B) \to C] \to [A \to (B \to C)]$, is true in M: Suppose there are $A, B, C \in \mathcal{F}$ and $a \in K$ in M such that (1) $a \vDash (A \wedge B) \to C$ but (2) $a \nvDash A \to (B \to C)$. Then, there are $b, c \in K$ in M such that (3) $Rabc$, (4) $b \vDash A$, and (5) $c \nvDash B \to C$, whence there are $d, e \in K$ such that (6) $Rcde$, (7) $d \vDash B$, and (8) $e \nvDash C$. By 3, 6, and d2, we get (9) R^2abde whence by pt31, there is some $x \in K$ in M such that (10) $Raxe$, (11) $b \leq x$, and (12) $d \leq x$. Next, we have (13) $x \vDash A \wedge B$, by 4, 7, 11, and 12; and, finally, (14) $e \vDash C$, by 1, 10, and 13. But 8 and 14 contradict each other.

(i) $t32$, $(A \to B) \vee [(A \to B) \to C]$, is true in M: Suppose there are $A, B \in \mathcal{F}$ and $a \in O$ in M such that (1) $a \nvDash (A \to B) \vee [(A \to B) \to C]$. Then, we have (2) $a \nvDash A \to B$ and (3) $a \nvDash (A \to B) \to C$, whence there are $b, c \in K$ in M such that (4) $Rabc$, (5) $b \vDash A$, and (6) $c \nvDash B$; and there are $d, e \in K$ in M such that (7) $Rade$, (8) $d \vDash A \to B$, and (9) $e \nvDash C$. As $a \in O$, we apply pt32 to 4 and 7 and we get (10) $Rdbc$. Finally, (11) $c \vDash B$ follows by 5, 8, and 10. But 6 and 11 contradict each other.

(j) $t37$, $(A \rightarrow C) \vee (B \rightarrow A)$, *is true in* M: Suppose there are $A, B \in \mathcal{F}$ and $a \in O$ in M such that (1) $a \nvDash (A \rightarrow C) \vee (B \rightarrow A)$. Then, we have (2) $a \nvDash A \rightarrow C$ and (3) $a \nvDash B \rightarrow A$, whence there are $b, c, d, e \in K$ in M such that (4) $Rabc$, (5) $b \vDash A$, (6) $c \nvDash C$, (7) $Rade$, (8) $d \vDash B$, and (9) $e \nvDash A$. Given that $a \in O$, pt37 is applied to 4 and 7, and we get (10) $b \leq e$. Finally, by 5 and 10, we have (11) $e \vDash A$, contradicting 9.

Proofs of t10, t11, t32, and t37 are developed for EB_+-models, but the proofs for EB_{K+}-models are similar since in EB_{K+}-models the set K has the properties that distinguish O from K in EB_+-models (the same remark applies to t22, t33, t34, t36–t38 that are proved with pt22, pt33, pt34, pt36–pt38, respectively). □

Next, we proceed to prove the adequacy of the semantical postulates. At this point, it may be useful to recall how canonical EB_+-models are defined. There are four types of canonical EB_+-models (cf. Definition 1.26). These are: (a) $(K^P, O^P, R^P, \vDash^P)$; (b) $(K^P, O^P, S^P, R^P, \vDash^P)$; (c) $(K^{NP}, O^P, S^{NP}, R^{NP}, \vDash^{NP})$; and (d) $(K^{WNP}, O^{WP}, R^{WNP}, \vDash^{WNP})$. On the other hand, there are three types of canonical B_{K+}-models (cf. Definition 3.18), which are the following: (a) $(K^{NP}, R^{NP}, \vDash^{NP})$; (b) $(K^{NP}, S^{NP}, R^{NP}, \vDash^{NP})$; and (c) $(K^{WNP}, R^{WNP}, \vDash^{WNP})$. In both classes of canonical models, the superscripts P, N, and W abbreviate prime, non-empty and w-consistent, respectively. Some postulates can be proved in any of the types of canonical models recalled above, but other postulates require non-empty and/or w-consistent theories in order to be proved. The propositions that follow classify postulates pt1–pt40 according to the requirements just mentioned.

Proposition 6.2 (Proof of pt1–pt23 in can. EB_+-models a and b). *Let L be an EB_+-logic and for any $j(1 \leq j \leq 23)$ let the canonical $L \cup \{tj\}$-model, M_C, be a canonical EB_+-model of type a or type b. Then, ptj is provable in M_C.*

Proof. The proof is similar to that provided in [37], Chapter 4, for extensions of Routley and Meyer's basic positive logic B_+, or, actually, to the proofs that postulates hold canonically, provided in preceding chapters. We prove Proposition 6.2 for the postulates used above in Proposition 6.1.

(a) *pt7*, $R^2 abcd \Rightarrow \exists x \exists y (Racx \mathbin{\&} Rbcy \mathbin{\&} Rxyd)$, *is provable in* M_C: Let $a, b, c, d \in K^P$ such that $R^{2P} abcd$. By d2, we have (1) $R^P abz$ and $R^P zcd$ for some $z \in K^P$. Consider now the sets $u = \{B \mid \exists A[A \rightarrow B \in a \mathbin{\&} A \in c]\}$

and $w = \{B \mid \exists A[A \to B \in b \ \& \ A \in c]\}$. By Lemma 1.24, u and w are L $\cup\{t7\}$-theories such that (2) $R^P acu$ and $R^P bcw$. Next, it is proved that $R^T uwd$ holds. Suppose, for $A, B \in \mathcal{F}$, (3) $A \to B \in u$ and (4) $A \in w$. We have to prove $B \in d$. By definitions of u and w, we have, for $C, C' \in c$, (5) $C \to (A \to B) \in a$, (6) $C' \to A \in b$. From 5 and 6, by using B_+, we have (7) $(C \wedge C') \to (A \to B) \in a$ and (8) $(C \wedge C') \to A \in b$. On the other hand, by t7, the following is a theorem of L $\cup\{t7\}$, (9) $[(C \wedge C') \to (A \to B)] \to [[(C \wedge C') \to A] \to [(C \wedge C') \to B]]$. Then, (10) $[(C \wedge C') \to A] \to [(C \wedge C') \to B] \in a$ follows by 7 and 9; and by 1, 8, and 10, we have (11) $(C \wedge C') \to B \in z$, whence we get (12) $B \in d$, by 1, since $C, C' \in c$, as it was required. Thus, we have $u, w \in K^T$ such that $R^T acu$, $R^T bcw$, and $R^T uwd$. Next, we extend u and w to the required prime L $\cup\{t7\}$-theories. By applying Lemmas 1.18 and 1.19 to $R^T uwd$, we have $x, y \in K^P$ such that $u \subseteq x$, $w \subseteq y$, and $R^P xyd$. Obviously, $R^P acx$ and $R^P bcy$. Therefore, we have $x, y \in K^P$ such that $R^P acx$, $R^P bcy$, and $R^P xyd$, as it was to be proved.

(b) *pt10*, $\exists x \in Z \ Raxa$ ($x \in Z$ *iff for any* $b, c \in K$, $Rxbc \Rightarrow \exists y \in O \ Rybc$), *is provable in* M_C: (Theorem t10' $[(A \to A) \to B] \to B$ will be used in the following proof. t10' is an easy consequence of t10 by using B_+.) Let $a \in K^P$. We have to prove that there is some $x \in K^P$ such that $R^P axa$ and $x \in Z$. Consider the set $u = \{A \mid \exists B[\vdash_{L \cup \{t10\}} (B \to B) \to A]\}$. We prove $u \in K^T$, i.e., that u is an L $\cup\{t10\}$-theory.

(bi) u is closed under L $\cup\{t10\}$-imp: Suppose, for $A, B \in \mathcal{F}$, (1) $\vdash_{L \cup \{t10\}} A \to B$ and (2) $A \in u$. By definition of u, we have, for some $C \in \mathcal{F}$, (3) $\vdash_{L \cup \{t10\}} (C \to C) \to A$. By 3 and Suf, we get (4) $\vdash_{L \cup \{t10\}} (A \to B) \to [(C \to C) \to B]$, and by 1 and 4, (5) $\vdash_{L \cup \{t10\}} (C \to C) \to B$, i.e., $B \in u$.

(bii) u is closed under Adj: Suppose, for $A, B \in \mathcal{F}$, (1) $A, B \in u$. By definition of u, we have (2) $\vdash_{L \cup \{t10\}} (C \to C) \to A$ and (3) $\vdash_{L \cup \{t10\}} (D \to D) \to B$. By 2, 3, and B_+, we get (4) $\vdash_{L \cup \{t10\}} [(C \to C) \wedge (D \to D)] \to (A \wedge B)$, whence by Suf we have (5) $\vdash_{L \cup \{t10\}} [(A \wedge B) \to (A \wedge B)] \to [[(C \to C) \wedge (D \to D)] \to (A \wedge B)]$. By applying again Suf, (6) $\vdash_{L \cup \{t10\}} [[[(C \to C) \wedge (D \to D)] \to (A \wedge B)] \to (A \wedge B)] \to [[(A \wedge B) \to (A \wedge B)] \to (A \wedge B)]$ is derivable from 5. Finally, we use t10 $[[(C \to C) \wedge (D \to D)] \to (A \wedge B)] \to (A \wedge B)$ to get (7) $\vdash_{L \cup \{t10\}} [(A \wedge B) \to (A \wedge B)] \to (A \wedge B)$ from 6. But 7 gives us (8) $A \wedge B \in u$, by definition of u, as it was to be proved.

Thus, by bi and bii, u is an L $\cup\{t10\}$-theory. Next, we prove that the relation $R^T aua$ obtains.

(biii) $R^T aua$: Suppose, for $A, B \in \mathcal{F}$, (1) $A \to B \in a$ and (2) $A \in u$. We have to prove $B \in a$. By definition of u, we get, for some $C \in \mathcal{F}$, (3) $\vdash_{L \cup \{t10\}} (C \to C) \to A$, and by Suf, (4) $\vdash_{L \cup \{t10\}} (A \to B) \to [(C \to$

$C) \to B]$. By 1 and 4, (5) $(C \to C) \to B \in a$ is derivable. Then, theorem t10′ remarked above is used to obtain (6) $B \in a$, as it was required.

It remains to extend u to a prime L $\cup\{t10\}$-theory x.

(biv) $R^P axa$: Given $R^T aua$, by Lemma 1.19, there is $x \in K^P$ such that $u \subseteq x$ and $R^P axa$.

Finally, we prove that x is a member of Z.

(bv) $x \in Z$: Suppose, for $b, c \in K^P$, (1) $R^P xbc$. We define the set z as the set of all theorems of L $\cup\{t10\}$: (2) $z = \{A \mid \vdash_{L \cup\{t10\}} A\}$. Suppose now for $A, B \in \mathcal{F}$, (3) $A \to B \in z$ and (4) $A \in b$. We have to prove $B \in c$. By 3 and definition of z, we get (5) $\vdash_{L \cup\{t10\}} A \to B$, whence by Suf (6) $\vdash_{L \cup\{t10\}}$ $(B \to B) \to (A \to B)$ is derivable, i.e., (7) $A \to B \in u$, by definition of u, and (8) $A \to B \in x$, since $u \subseteq x$. Finally, we obtain (9) $B \in c$, by 1, 4, and 8, as it was required. It remains to extend z to a prime L $\cup\{t10\}$-theory $y \in O$ such that $R^P ybc$, which is immediate by applying Lemma 1.18 to $R^T zbc$.

(c) $pt11$, $\exists x \in O\, Raxa$, is provable in M_C: Let $a \in K^P$ and y be the set of all theorems of L $\cup\{t11\}$, i.e., $y = \{A \mid \vdash_{L \cup\{t11\}} A\}$. We prove $R^T aya$. Suppose, for $A, B \in \mathcal{F}$, (1) $A \to B \in a$ and (2) $A \in y$. We have to prove $B \in a$. By 2 and definition of y, (3) $\vdash_{L \cup\{t11\}} A$ follows, whence, by t11, we have (4) $\vdash_{L \cup\{t11\}}$ $(A \to B) \to B$ and finally (5) $B \in a$, by 1 and 4. Next, by Lemma 1.19, there is some $x \in K^P$ such that (6) $R^P axa$. Obviously, $x \in O^P$.

(d) $pt13$, $R^3 abcde \Rightarrow R^3 acbde$, is provable in M_C: Suppose there are $a, b, c, d, e \in K^P$ such that $R^{3P} abcde$. By d3, for $x, y \in K^P$, we have (1) $R^P abx$, $R^P xcy$ and $R^P yde$. Consider now the sets $z = \{B \mid \exists A[A \to B \in a \;\&\; A \in c]\}$ and $u = \{B \mid \exists A[A \to B \in z \;\&\; A \in b]\}$. By Lemma 1.24 z and u are L $\cup\{t13\}$-theories such that (2) $R^T acz$ and $R^T zbu$. Next, it is proved that $R^T ude$ holds. Suppose, for $A, B \in \mathcal{F}$, (3) $A \to B \in u$ and (4) $A \in d$. We have to prove $B \in e$. By definitions of z and u, we get (5) $D \to$ $[C \to (A \to B)] \in a$ for $C \in b$ and $D \in c$. By t13, (6) $[D \to [C \to (A \to$ $B)]] \to [C \to [D \to (A \to B)]]$ is a theorem of L $\cup\{t13\}$. Then, we have: (7) $C \to [D \to (A \to B)] \in a$ (by 5 and 6), (8) $D \to (A \to B) \in x$ (by 1 and 7, as $C \in b$), (9) $A \to B \in y$ (by 1 and 8, since $D \in c$) and, finally, (10) $B \in e$ (by 1, 4, and 9), as it was to be proved. Thus, we have $z, u \in K^T$ such that $R^T acz, R^T zbu$ and $R^T ude$. It remains to extend u and z to the required prime L $\cup\{t13\}$-theories. By Lemma 1.18, u is extended to $v \in K^P$ such that $R^P vde$; clearly, $R^T zbv$. Next, by applying again Lemma 1.18, z is extended to $w \in K^P$ such that $R^P wbv$; clearly, $R^P acw$. Consequently, we have $v, w \in K^P$ such that $R^P acw, R^P wbv$ and $R^P vde$, i.e., $R^{3P} acbde$, as it was to be proved.

(e) *pt20*, $Rabc \Rightarrow (a \leq c$ *or* $b \leq c)$, *is provable in* M_C: (The thesis t21, $A \to [B \to (A \vee B)]$, an easy consequence of t20, $A \to (A \to A)$, by B_+ will be used in the proof that follows.) Suppose (1) $R^P abc$ for $a, b, c \in K^P$. And, for reductio, suppose (2) $a \not\leq^P c$ and $b \not\leq^P c$. By Corollary 1.23 and 2, there are $A, B \in \mathcal{F}$ such that (3) $A \in a$, $A \notin c$, $B \in b$, and $B \notin c$. By t21, we get (4) $A \to [B \to (A \vee B)]$. Then, we have the following: (5) $B \to (A \vee B) \in a$ (by 3 and 4) and (6) $A \vee B \in c$ (by 1, 3, and 5). But 3 and 6 contradict each other since c is prime.   □

Proposition 6.3 (Proof of pt1–pt31 in can. EB_{K+}-models a and b). *Let L be an EB_{K+}-logic and for any $j(1 \leq j \leq 31)$ let the canonical L $\cup\{tj\}$-model, M_C, be a canonical EB_{K+}-model of type a or type b. Then, ptj is provable in M_C.*

Proof. (I) *pt1–pt23*. pt1–pt23 are proved exactly as in Proposition 6.2 except that each new theory introduced has to be shown non-empty. Thus, for example, in pt1, the required new theory x has to be proved non-empty. But this is easily accomplished by using Lemma 3.17 Consider, for example, pt13 as proved in Proposition 6.2. There it is shown that given $R^{3P} abcde$ for $a, b, c, d, e \in K^P$, there are $w, w' \in K^P$ such that $R^P acw$, $R^P wbw'$, and $R^P w' de$. Let now L $\cup\{t13\}$ be an EB_{K+}-logic and the canonical L $\cup\{t13\}$-model be a canonical EB_{K+}-model of type a or b; and let $R^{3NP} abcde$ for $a, b, c, d, e \in K^{NP}$. Exactly as in Proposition 6.2, we prove that there are $w, w' \in K^P$ such that $R^P acw$, $R^P wbw'$, and $R^P w' de$. Now, by applying Lemma 3.17 to $R^P acw$ it follows that $w \in K^{NP}$; then by applying the same lemma to $R^P wbw'$, we get that w' is also in K^{NP}. (By the way, recall that any EB_{K+}-theory is non-empty iff it is regular; cf. Lemma 3.12.)

(II) *pt24–pt31*. Concerning pt24–pt31, we proceed similarly as in Proposition 6.2. Let us prove some examples.

(a) *pt24*, $(Rabc$ & $Rade) \Rightarrow \exists x[(Raxc$ *or* $Raxe)$ & $b \leq x$ & $d \leq x]$, *is provable in* M_C: Let $a, b, c, d, e \in K^{NP}$ such that (1) $R^{NP} abc$ and $R^{NP} ade$. Consider the set $y = \{C \mid \exists A, B[\vdash_{L \cup\{t24\}} (A \wedge B) \to C$ & $A \in b$ & $B \in d]\}$. The resources of B_+ suffice to prove easily that y is an L $\cup\{t24\}$-theory. On the other hand, it is obvious that b and d are included in y: if $A \in b$, $B \in d$, then $A \in y$ and $B \in y$ by A2 of B_+ $((A \wedge B) \to A; (A \wedge B) \to B)$. Thus, notice that y is non-empty. So it remains to prove $R^{NT} ayc$ or $R^{NT} aye$. Suppose, for reductio, (2) Not-$R^{NT} ayc$ and Not-$R^{NT} aye$. By definition of y, we have, for $C, C', D, D' \in \mathcal{F}$, (3) $C \to D \in a$, $C \in y$, and $D \notin c$ and (4) $C' \to D' \in a$, $C' \in y$, and $D' \notin e$. By 3, 4, and B_+, we get (5) $(C \wedge C') \to (D \wedge D') \in a$; and by 3, 4, and definition of y, (6) $\vdash_{L \cup\{t24\}} (A \wedge B) \to (C \wedge C')$ for $A \in b$ and $B \in d$. Then, by 5 and 6, (7) $(A \wedge B) \to (D \wedge D') \in a$ follows. Now, by t24, (8) $[(A \wedge B) \to (D \wedge D')] \to [[A \to (D \wedge D')] \vee [B \to (D \wedge D')]]$ is a theorem. So, we have (9) $[A \to (D \wedge D')] \vee [B \to (D \wedge D')] \in a$ by 7 and 8. Suppose now (10) $A \to (D \wedge D') \in a$. Then, we have (11) $D \wedge D' \in c$ (by 1 since $A \in b$). But 11 contradicts 3. Suppose, on the other hand, (12) $B \to (D \wedge D') \in a$. Then, we have (13) $D \wedge D' \in e$ (by 1 since $B \in d$). But 13 contradicts 4. Consequently, we conclude (14) $R^{NT}ayc$ or $R^{NT}aye$. Finally, by Lemma 1.19, there is some $x \in K^P$ such that $y \subseteq x$ and $R^{NP}axc$ or $R^{NP}axe$. Clearly, $b \subseteq x$ and $d \subseteq x$ (as $b \subseteq y$ and $d \subseteq y$), whence by Corollary 3.16, we have $b \leq^P x$ and $d \leq^P x$, which completes the proof that pt24 is provable in M_C.

(b) *pt26, $R^2abcd \Rightarrow Racd$, is provable in M_C*: Let $a, b, c, d \in K^{NP}$ such that $R^{2NP}abcd$. By d2, we have (1) $R^{NP}aby$ and $R^{NP}ycd$ for some $y \in K^{NP}$. For $A, B \in \mathcal{F}$, suppose (2) $A \to B \in a$ and (3) $A \in c$. We have to prove $B \in d$. Suppose (4) $C \in b$. By t26, (5) $(A \to B) \to [C \to (A \to B)]$ is a theorem. So, (6) $C \to (A \to B) \in a$ follows (by 2 and 5), whence we get (7) $A \to B \in y$, by 1, 4, and 6. Finally, we have (8) $B \in d$, by 1, 3, and 7, as it was to be proved.

(c) *pt27, $Rabc \Rightarrow a \leq c$, is provable in M_C*: Let $a, b, c \in K^{NP}$ such that (1) $R^{NP}abc$ and suppose (2) $A \in a$ and $B \in b$. By t27, we have (3)$\vdash_{L \cup\{t27\}} A \to (B \to A)$. Thus, we have the following: (4) $B \to A \in a$ (by 2 and 3) and (5) $A \in c$ (by 1, 2, and 4). So, we have (6) $a \subseteq c$, whence (7) $a \leq^P c$ follows by Corollary 3.16.

(d) *pt31, $R^2abcd \Rightarrow \exists x(Raxd \ \& \ b \leq x \ \& \ c \leq x)$, is provable in M_C*: Let $a, b, c, d \in K^{NP}$ such that $R^{2NP}abcd$. By d2, we have (1) $R^{NP}aby$ and $R^{NP}ycd$ for some $y \in K^{NP}$. Consider the set $z = \{C \mid \exists A, B[\vdash_{L \cup\{t31\}} (A \wedge B) \to C \ \& \ A \in b \ \& \ B \in c]\}$. Similarly as in case (a), it is proved that z is an L $\cup\{t31\}$-theory such that (2) $b \subseteq z$ and $c \subseteq z$. Moreover, we clearly have (3) z is non-empty. It remains to prove $R^{NP}azd$ and to extend z to the required prime L $\cup\{t31\}$-theory. Then, suppose (4) $A \to B \in a$ and (5) $A \in z$ for $A, B \in \mathcal{F}$. We have to prove $B \in d$. By 5 and definition of z, we get (6) $\vdash_{L \cup\{t31\}} (C \wedge D) \to A$ for $C \in b$ and $D \in c$, whence (7) $(C \wedge D) \to B \in a$ follows by 4. Now, by t31, (8) $[(C \wedge D) \to B] \to [C \to (D \to B)]$ is a theorem. So (9) $C \to (D \to B) \in a$. Then, we have the following: (10) $D \to B \in y$ (by 1 and 9 since $C \in b$) and finally (11) $B \in d$ (by 1 and 10 since $D \in c$), as it was required. Next, z is extended to the required prime L $\cup\{t31\}$-theory. By Lemma 1.19, there is some $x \in K^P$ such that $z \subseteq x$ and $R^{NP}axd$. Obviously $b \subseteq x$ and $c \subseteq x$ (since by 2, $b \subseteq z$ and $c \subseteq z$). Thus, $b \leq^P x$ and $c \leq^P x$ by Corollary 3.16. $\square$

Proposition 6.4 (Proof of pt1–pt31 in can. EB_+-models c). *Let L be an EB_{cS}-logic and for any $j(1 \leq j \leq 31)$ let the canonical $L \cup \{tj\}$-model, M_C, be a canonical EB_+-model of type c. Then, ptj is provable in M_C.*

Proof. It is similar to that of the preceding proposition. Postulates pt1–pt23 are proved similarly as in Proposition 6.2, except that non-emptiness of new introduced theories either follows from their definition (as it is the case with pt10, pt11, pt23, pt24, and pt31) or else by Lemma 2.15, as it holds with the rest of the postulates (cf. Proposition 6.3). Consider, for example, pt10, as it is proved in Proposition 6.2. The theory x is an extension of the theory u that is defined as follows: $u = \{A \mid \exists B[\vdash_{L \cup \{t10\}} (B \to B) \to A]\}$. But u is obviously non-empty: by A1, $B \to B \in u$. Or, to take another example, consider t24 as it is proved in the same Proposition 6.3. Again, x is an extension of y, which is defined as follows: $y = \{C \mid \exists A, B[\vdash_{L \cup \{t24\}} (A \wedge B) \to C \ \& \ A \in b \ \& \ B \in d]\}$. Suppose that b and d are non-empty and $A \in b$, $B \in d$: by A2, $(A \wedge B) \to A$, $A \in y$ follows.

On the other hand, pt24–pt31 are proved similarly as in Proposition 6.3. Consider, for example, pt25 $Rabc \Rightarrow b \leq c$. Suppose for $A, B \in \mathcal{F}$ and $a, b, c \in K^{NP}$, (1) $R^{NP}abc$, (2) $B \in a$, and (3) $A \in b$. By t25, (4) $B \to (A \to A)$ is a theorem. Then, we have (5) $A \to A \in a$ (by 2 and 4) and, finally, (6) $A \in c$ (by 1, 3, and 5). That is, (7) $b \subseteq c$, whence (8) $b \leq^P c$ by Corollary 1.23, as it was to be proved. $\square$

Proposition 6.5 (Pr. of pt1–pt40 in can. EB_+-m. d and EB_{K+}-m. c). *Let L be an EB_c-logic and for any $j(1 \leq j \leq 40)$ let the canonical $L \cup \{tj\}$-model, M_C, be a canonical EB_+-model of type d or a canonical EB_{K+}-model of type c. Then, ptj is provable in M_C.*

Proof. (I) *pt1–pt31.* Postulates pt1–pt31 are proved similarly as in Proposition 6.3 or Proposition 6.4 except that each new theory introduced has to be shown w-consistent. But the w-consistency of the new introduced theories follows from Lemma 2.17. Let us illustrate this fact with an example.

pt7, $R^2abcd \Rightarrow \exists x, y(Racx \ \& \ Rbcy \ \& \ Rxyd)$, *is provable in* M_C: Leaning on Proposition 6.3 or Proposition 6.4 and given $R^{WNP}abcd$ for $a, b, c, d \in K^{WNP}$, we have $x, y \in K^{NP}$ such that $R^{NP}acx$, $R^{NP}bcy$, and $R^{NP}xyd$. Then, $R^{WNP}acx$, $R^{WNP}bcy$, and $R^{WNP}xyd$ immediately follow by Lemma 2.17.

(II) *pt32–pt40.* On the other hand, pt32–pt40 are proved on the basis that all elements of the canonical set K are w-consistent. Let us prove a couple of examples, the first one for canonical EB_+-models, and the second one for canonical EB_{K+}-models.

pt32, (Rabc & Rade & $a \in O$) $\Rightarrow$ Rdbc, is provable in M_C: Let $a \in O^{\mathrm{WP}}$ and $b, c, d, e \in K^{\mathrm{WNP}}$ such that (1) $R^{\mathrm{WNP}}abc$ and $R^{\mathrm{WNP}}ade$. Further, suppose (2) $A \to B \in d$ and (3) $A \in b$ for $A, B \in \mathcal{F}$. We have to prove $B \in c$. Suppose, for reductio, (4) $B \notin c$. We have (5) $A \to B \notin a$ since $R^{\mathrm{WNP}}abc$, $A \in b$ and $B \notin c$ by 1, 3, and 4. Now, by t32, (6) $(A \to B) \lor [(A \to B) \to C]$ is a theorem of L $\cup\{$t32$\}$, C being an arbitrary formula. So, (7) $A \to B \in a$ or $(A \to B) \to C \in a$ follows, as $a \in O^{\mathrm{WP}}$. Consequently, we get (8) $(A \to B) \to C \in a$, whence we have (9) $C \in e$, by 1 and 2. But 9 contradicts the w-consistency of e.

pt39, (Rabc & Rade) $\Rightarrow$ ($a \leq c$ or $b \leq e$), is provable in M_C: Suppose for $a, b, c, d, e \in K^{\mathrm{WNP}}$ (1) $R^{\mathrm{WNP}}abc$ and $R^{\mathrm{WNP}}ade$, and, for reductio, (2) $A \in a$, $A \notin e$, $B \in b$, and $B \notin e$ for some $A, B \in \mathcal{F}$. Moreover, let (3) $D \in d$ and C be the negation of a theorem of L $\cup\{$t39$\}$. We have (4) $A \lor B \in a$ (by 2 and A4); (5) $A \lor B \in b$ (by 2 and A4); (6) $A \lor B \notin e$ (by 2); (7) $(A \lor B) \to C \notin a$ (by 1, 3, and 5: $R^{\mathrm{WNP}}abc$, $A \lor B \in b$, and $C \notin c$); and (8) $D \to (A \lor B) \notin a$ (by 1, 3, and 6: $R^{\mathrm{WNP}}ade$, $D \in d$, $A \lor B \notin e$). By t39, (9) $(A \lor B) \to [[(A \lor B) \to C] \lor [D \to (A \lor B)]]$ is a theorem of L $\cup\{$t39$\}$. So, (10) $[(A \lor B) \to C] \lor [D \to (A \lor B)] \in a$ follows by 4 and 9. But 7 and 8 contradict 10. $\square$

Thus, we see, pt1–pt23 can be proved in any of the types of canonical models we have defined (cf. Definitions 1.26 and 3.18), but pt24–pt31 demand non-empty theories; pt32–pt36, w-consistent theories, and pt37–pt40 both non-empty and w-consistent theories. Consequently, any EB_+-logic can be extended with t1–t23, but only EB_{cS}-logics or EB_{K+}-logics can be strengthened with t24–t31, whereas t32–t40 require at least EB_c-logics. (Notice that Lemma 2.17 requires non-empty theories.)

We remark that the requisite of w-consistency in pt32–pt40 can be changed by alternative notions of consistency such as u-consistency, a-consistency (cf. Definition 1.15) or even the standard notion of consistency in function of the logic L in question. Suppose, for example, that S is interpreted as the set of all a-consistent theories. Then, pt32–pt36 can be proved in any EB_+-logic L in which the corresponding thesis holds, provided the condition $c \in S$ (respectively, $e \in S$) be added to pt33, pt35, and pt36 (respectively, pt32 and pt34). On the other hand, pt37–pt40 can be proved in any EB_{K+}-logic L where the corresponding thesis holds, assuming that the condition $c \in S$ (respectively, $e \in S$) be added to pt37, pt39, and pt40 (respectively, pt38). But we cannot discuss this question in the present work.

6.2 EXTENSIONS AND EXPANSIONS WITH NEGATION AXIOMS

Consider the following theses:

List of negation axioms:

t41. $[(A \to B) \land \neg B] \to \neg A$

t42. $\neg B \to [(A \to B) \to \neg A]$

t43. $(A \to B) \to (\neg B \to \neg A)$

t44. $(A \to \neg B) \to (B \to \neg A)$

t45. $B \to [(A \to \neg B) \to \neg A]$

t46. $A \to [(A \to \neg B) \to \neg B]$

t47. $[A \to (B \to \neg C)] \to [B \to (A \to \neg C)]$

t48. $A \to \neg\neg A$

t49. $\neg(A \land \neg A)$

t50. $(A \to \neg A) \to \neg A$

t51. $A \to \neg(A \to \neg A)$

t52. $(A \to B) \to \neg(A \land \neg B)$

t53. $(A \land \neg B) \to \neg(A \to B)$

t54. $(A \to \neg B) \to \neg(A \land B)$

t55. $(A \land B) \to \neg(A \to \neg B)$

t56. $(A \to \neg B) \to [(A \to B) \to \neg A]$

t57. $(A \to B) \to [(A \to \neg B) \to \neg A]$

t58. $\neg(A \land B) \to (\neg A \lor \neg B)$

t59. $\neg A \lor \neg\neg A$

t60. $(A \to B) \lor \neg(A \to B)$

t61. $A \lor \neg A$

t62. $A \to [B \lor \neg(A \to B)]$

t63. $A \lor [A \to (B \lor \neg B)]$

t64. $(A \lor \neg B) \lor (A \to B)$

t65. $B \to (A \lor \neg A)$

t66. $\neg A \lor (B \to A)$

t67. $A \lor (B \to \neg A)$

t68. $\neg A \to (B \to \neg A)$

t69. $B \rightarrow [\neg B \vee (A \rightarrow B)]$

t70. $(A \wedge \neg A) \rightarrow B$

t71. $A \rightarrow (\neg A \rightarrow B)$

Consider now the following semantical postulates corresponding to t40–t71:

List of negation postulates:

pt41. $(Rabc \ \& \ c \in S) \Rightarrow \exists x \in K \ \exists y \in S(Rabx \ \& \ Raxy)$

pt42. $(R^2 abcd \ \& \ d \in S) \Rightarrow \exists x \in K \ \exists y \in S(Rbcx \ \& \ Raxy)$

pt43. $(R^2 abcd \ \& \ d \in S) \Rightarrow \exists x \in K \ \exists y \in S(Racx \ \& \ Rbxy)$

pt44. $(R^2 abcd \ \& \ d \in S) \Rightarrow \exists x \in S \ R^2 acbx$

pt45. $(R^2 abcd \ \& \ d \in S) \Rightarrow \exists x \in S \ R^2 bcax$

pt46. $(R^2 abcd \ \& \ d \in S) \Rightarrow \exists x \in S \ R^2 bacx$

pt47. $(R^3 abcde \ \& \ e \in S) \Rightarrow \exists x \in S \ R^3 acbdx$

pt48. $(Rabc \ \& \ c \in S) \Rightarrow \exists x \in S \ Rbax$

pt49. $(Rabc \ \& \ a \in O \ \& \ c \in S) \Rightarrow \exists x \in S \ Rbbx$

pt50. $(Rabc \ \& \ c \in S) \Rightarrow \exists x \in S \ R^2 abbx$

pt51. $(Rabc \ \& \ c \in S) \Rightarrow \exists x \in S \ R^2 baax$

pt52. $(Rabc \ \& \ c \in S) \Rightarrow \exists x \in K \ \exists y \in S(Rabx \ \& \ Rbxy)$

pt53. $(Rabc \ \& \ c \in S) \Rightarrow \exists x \in K \ \exists y \in S(Rbax \ \& \ Raxy)$

pt54. $(Rabc \ \& \ c \in S) \Rightarrow \exists x \in S \ R^2 abbx$

pt55. $(Rabc \ \& \ c \in S) \Rightarrow \exists x \in S \ R^2 baax$

pt56. $(R^2 abcd \ \& \ d \in S) \Rightarrow \exists x, y \in K \ \exists z \in S(Racx \ \& \ Rbcy \ \& \ Rxyz)$

pt57. $(R^2 abcd \ \& \ d \in S) \Rightarrow \exists x, y \in K \ \exists z \in S(Racx \ \& \ Rbcy \ \& \ Ryxz)$

pt58. $(Rabc \ \& \ Rade \ \& \ c \in S \ \& \ e \in S) \Rightarrow \exists x \in K \ \exists y \in S(Raxy \ \&$
$b \leq x \ \& \ d \leq x)$

pt59. $(Rabc \ \& \ Rade \ \& \ a \in O \ \& \ c \in S \ \& \ e \in S) \Rightarrow \exists x \in S \ Rdbx$

pt60. $(Rabc \ \& \ Rade \ \& \ a \in O \ \& \ e \in S) \Rightarrow \exists x \in S(Rdbx \ \& \ x \leq c)$

pt61. $(Rabc \ \& \ a \in O \ \& \ c \in S) \Rightarrow b \leq a$

pt62. $(Rabc \ \& \ c \in S) \Rightarrow \exists x \in K(Rbax \ \& \ x \leq a)$

pt63. $(Rabc \ \& \ Rcde \ \& \ a \in O \ \& \ e \in S) \Rightarrow (b \leq a \ \text{or} \ d \leq c)$

pt64. $(Rabc \ \& \ Rade \ \& \ a \in O \ \& \ c \in S) \Rightarrow (d \leq a \ \text{or} \ b \leq e)$

pt65. $(Rabc \ \& \ c \in S) \Rightarrow b \leq a$

pt66. $(Rabc \ \& \ Rade \ \& \ a \in O \ \& \ c \in S) \Rightarrow b \leq e$

pt67. $(Rabc \ \& \ Rcde \ \& \ a \in O \ \& \ e \in S) \Rightarrow d \leq a$

pt68. $(R^2abcde \ \& \ d \in S) \Rightarrow \exists x \in S \ Racx$

pt69. $(Rabc \ \& \ Rade \ \& \ c \in S) \Rightarrow (a \leq e \text{ or } b \leq e)$

pt70. $\exists x \in K \ Raax$

pt71. $Rabc \Rightarrow \exists x \in K \ Rbax$

We shall define extensions of the basic logics by using t41–t71 and RM-semantics for these extensions by means of the corresponding postulates pt41–pt71. We note that t44, t48, t50, t61, t65, and t71 are among the theses considered in [37] for extending Routley and Meyer's basic logic B.

We begin by proving the validity of t41–t71 in any EB_+-model or EB_{K+}-model in which the corresponding semantical postulate holds.

Proposition 6.6 (Validity of t41–t71). *Let a class of EB_+-models or EB_{K+}-models, $\mathcal{M}$, be defined and $M \in \mathcal{M}$. Then, for any j ($41 \leq j \leq 71$), tj is true in M if ptj holds in M.*

Proof. We remark that the validity of t44 and t45 has been proved in Proposition 5.3. The rest of the theses is proved in a similar way (cf. the comment at the beginning of the proof of Proposition 6.1). We prove a few cases. We use clause (v) in the first five cases, and clause (vi) in the two last ones.

(a) *t46, $A \rightarrow [(A \rightarrow \neg B) \rightarrow \neg B]$, is true in M*: Suppose there are $A, B \in \mathcal{F}$ and $a \in K$ in M such that (1) $a \vDash A$ but (2) $a \nvDash (A \rightarrow \neg B) \rightarrow \neg B$. Then, there are $b, c \in K$ such that (3) $Rabc$, (4) $b \vDash A \rightarrow \neg B$, and (5) $c \nvDash \neg B$. By 5 and v, there are $d \in K$ and $e \in S$ such that (6) $Rcde$ and (7) $d \vDash B$. By 3, 6, and d2, we have (8) R^2abde, whence by pt46 and d2 it follows that there are $y \in K$ and $x \in S$ such that (9) $Rbay$ and (10) $Rydx$. Then, we have (11) $y \vDash \neg B$ (by 1, 4, and 9) and, finally, (12) $d \nvDash B$ (by 10 and 11, since $x \in S$). But 7 and 12 contradict each other.

(b) *t58, $\neg(A \wedge B) \rightarrow (\neg A \vee \neg B)$, is true in M*: Suppose there are $A, B \in \mathcal{F}$ and $a \in K$ in M such that (1) $a \vDash \neg(A \wedge B)$ but (2) $a \nvDash \neg A \vee \neg B$. Then, (3) $a \nvDash \neg A$ and (4) $a \nvDash \neg B$ follow. By 3 and v, we get (5) $Rabc$ and (6) $b \vDash A$, for $b \in K$ and $c \in S$; by 4 and v, we get (7) $Rade$ and (8) $d \vDash B$ for $d \in K$ and $e \in S$. Next, we apply pt58 to 5 and 7 (as $c, e \in S$), and we have (9) $Raxy$, (10) $b \leq x$, and (11) $d \leq x$ for $x \in K$ and $y \in S$. Then, by 6, 8, 10, and 11, we get (12) $x \vDash A \wedge B$. But, on the other hand, by 1, 9, and v, given $y \in S$, we have (13) $x \nvDash A \wedge B$, a contradiction.

(c) *t59*, $\neg A \vee \neg\neg A$, *is true in* M: Suppose, for $A \in \mathcal{F}$ and $a \in O$ in M, (1) $a \nvDash \neg A \vee \neg\neg A$. Then (2) $a \nvDash \neg A$ and (3) $a \nvDash \neg\neg A$ follow. By 2 and v we get, for $b \in K$ and $c \in S$, (4) *Rabc* and (5) $b \vDash A$; by 3 and v, we get, for $d \in K$ and $e \in S$, (6) *Rade* and (7) $d \vDash \neg A$. By pt59, 4, and 6, (as $a \in O$ and $c, e \in S$), we have (8) *Rdbx* for $x \in S$, whence we derive (9) $b \nvDash A$ by 7, contradicting 5.

(d) *t64*, $(A \vee \neg B) \vee (A \to B)$, *is true in* M: Suppose, for $A, B \in \mathcal{F}$ and $a \in O$ in M, (1) $a \nvDash A$, (2) $a \nvDash \neg B$, and (3) $a \nvDash A \to B$. By 2, there are $b \in K$ and $c \in S$ such that (4) *Rabc* and (5) $b \vDash B$; by 3, there are $d, e \in K$ such that (6) *Rade*, (7) $d \vDash A$, and (8) $e \nvDash B$. Given 4, 6 and the facts $a \in O$ and $c, e \in S$, pt64 is applicable and we have (9) $d \leq a$ or (10) $b \leq e$. But, if 9 is the case, then by 7, we have (11) $a \vDash A$, contradicting 1. On the other hand, if 10 obtains, then by 5, we have (12) $e \vDash B$, contradicting 8.

(e) *t69*, $B \to [\neg B \vee (A \to B)]$, *is true in* M: Suppose there are $A, B \in \mathcal{F}$ and $a \in K$ in M such that (1) $a \vDash B$ but (2) $a \nvDash \neg B$ and (3) $a \nvDash (A \to B)$. By 2, there are $b \in K$ and $c \in S$ such that (4) *Rabc* and (5) $b \vDash B$; by 3 there are $d, e \in K$ such that (6) *Rade*, (7) $d \vDash A$, and (8) $e \nvDash B$. Next, we apply pt69 to 4 and 6 (as $c \in S$) and we obtain (9) $a \leq e$ or (10) $b \leq e$, whence, as in the preceding case, a contradiction arises by using 1, 5, and 8.

As pointed out above, we use clause (vi) in the two cases that follow. We remark that clause (v) is insufficient for treating these cases.

(f) *t70*, $(A \wedge \neg A) \to B$, *is true in* M: Suppose there are $A, B \in \mathcal{F}$ and $a \in K$ in M such that (1) $a \vDash A$, (2) $a \vDash \neg A$ but (3) $a \nvDash B$. By pt70, we have (4) *Raax*, for some $x \in K$; and by 2 and 4, we get (5) $a \nvDash A$, contradicting 1.

(g) *t71*, $A \to (\neg A \to B)$, *is true in* M: Suppose there are $A, B \in \mathcal{F}$ and $a \in K$ in M such that (1) $a \vDash A$, (2) $a \nvDash \neg A \to B$. Then, there are $b, c \in K$ such that (3) *Rabc*, (4) $b \vDash \neg A$, and (5) $c \nvDash B$. By pt71, we have (6) *Rbax*, for some $x \in K$; and by 4 and 6, we get (7) $a \nvDash A$, contradicting 1.

The proofs of the validity of t59 and t64 are developed for EB_+-models. However, the proof of these theses in EB_{K+}-models is similar, as it has been pointed out more than once above (the same remark would apply to theses t49, t60, t61, t63, t66, and t67). $\qquad\qquad\square$

Next, we turn to the proof of the adequacy of the semantical postulates pt41–pt71 for t41–t71. Recall that we consider four types of canonical EB_+-models (cf. Definition 1.26) and three types of EB_{K+}-models (cf. Definition 3.18). The propositions that follow classify pt41–pt71, according to the requirements to prove them in these types of canonical models.

Proposition 6.7 (Proof of pt41–pt64 in can. EB_+-models b). *Let L be an ERB_c-logic and for any $j(41 \leq j \leq 64)$ let the canonical $L \cup \{tj\}$-model, M_C, be a canonical EB_+-model of type b. Then, ptj is provable in M_C.*

Proof. We remark that pt44 and pt45 have been proved in Lemma 5.6. The rest of the postulates is proved in a similar way. We prove a few cases (we use T4 of $\mathcal{R}B_c$ also labelled A7' —cf. Lemma 2.21— or A9 —cf. Definition 3.22).

(a) pt52, $(Rabc \ \& \ c \in S) \Rightarrow \exists x \in K \ \exists y \in S(Rabx \ \& \ Rbxy)$, *is provable in M_C*: Let $a, b \in K^P$, $c \in S^P$ and (1) $R^P abc$. Consider the set $z = \{B \mid \exists A[A \rightarrow B \in a \ \& \ A \in b]\}$. By Lemma 1.24, $z \in K^T$ and (2) $R^T abz$. Consider now the set $u = \{B \mid \exists A[A \rightarrow B \in b \ \& \ A \in z]\}$. By applying again Lemma 1.24, we have (3) $u \in K^T$ and $R^T bzu$. Next, we prove that u is w-consistent. Suppose it is not. Then, (4) $\neg A \in u$, A being some theorem of $L \cup \{t52\}$. By definitions of z and u, we have, for some $B, C \in \mathcal{F}$, (5) $B \rightarrow \neg A \in b$, (6) $C \rightarrow B \in a$, and (7) $C \in b$. By t52, (8) $(C \rightarrow B) \rightarrow \neg(C \wedge \neg B)$ is a theorem. So, by 6 and 8, (9) $\neg(C \wedge \neg B) \in a$ follows. On the other hand, by 5 and the rule Red, we have (10) $\neg B \in b$, whence by 7, we get (11) $C \wedge \neg B \in b$. Next, we apply A7' (12) $\neg(C \wedge \neg B) \rightarrow [(C \wedge \neg B) \rightarrow \neg[(C \wedge \neg B) \rightarrow (C \wedge \neg B)]]$. Then, we have the following: (13) $(C \wedge \neg B) \rightarrow \neg[(C \wedge \neg B) \rightarrow (C \wedge \neg B)] \in a$, by 9 and 12; (14) $\neg[(C \wedge \neg B) \rightarrow (C \wedge \neg B)] \in c$, by 1, 11, and 13. But 14 contradicts the w-consistency of c. Consequently, u is w-consistent. Finally, z and u are extended to the required $x \in K^P$ and $y \in S^P$ such that $R^P abx$, $R^P bxy$. This is done in the customary way by using Lemmas 2.13 and 1.19.

(b) pt58, $(Rabc \ \& \ Rade \ \& \ c \in S \ \& \ e \in S) \Rightarrow \exists x \in K \ \exists y \in S(Raxy \ \& \ b \leq x \ \& \ d \leq x)$, *is provable in M_C*: Let $a, b, d \in K^P$, $c, e \in S^P$ and (1) $R^P abc$ and (2) $R^P ade$. Consider the set $z = \{B \mid \forall A \in \mathcal{F}[\vdash_{L \cup \{t58\}} A \Rightarrow (B \rightarrow \neg A \notin a)]\}$. We prove that z is an $L \cup \{t58\}$-theory.

(bi) z is closed under $L \cup \{t58\}$-imp: Suppose for $B, C \in \mathcal{F}$, (3) $\vdash_{L \cup \{t58\}} B \rightarrow C$ and (4) $B \in z$. We have to prove $C \in z$. By 4 and definition of z we have (5) $B \rightarrow \neg A \notin a$, for any theorem A of $L \cup \{t58\}$. Suppose now for reductio (6) $C \notin z$, i.e., $C \rightarrow \neg D \in a$, D being a theorem of $L \cup \{t58\}$. By the rule Red, we get (7) $\neg C \in a$; by 3 and the rule Con, (8) $\vdash_{L \cup \{t58\}} \neg C \rightarrow \neg B$, and by 7 and 8, (9) $\neg B \in a$. Next, we apply A7' (10) $\neg B \rightarrow [B \rightarrow \neg(B \rightarrow B)]$. By 9 and 10, we get (11) $B \rightarrow \neg(B \rightarrow B) \in a$, whence, by definition of z, we have (12) $B \notin z$, contradicting 4. Therefore, $C \in z$, as it was to be proved.

(bii) z is closed under Adj: Suppose for $B, C \in \mathcal{F}$, (13) $B \in z$ and (14) $C \in z$. By 13, 14, and definition of z, for any theorem A of $L \cup \{t58\}$, we have (15) $B \to \neg A \notin a$ and (16) $C \to \neg A \notin a$. Suppose for reductio (17) $B \wedge C \notin z$, i.e., (18) $(B \wedge C) \to \neg D \in a$, D being a theorem of L $\cup\{t58\}$. By Red, we have (19) $\neg(B \wedge C) \in a$, whence by t58, we get (20) $\neg B \vee \neg C \in a$. By primeness of a, either (21) $\neg B \in a$ or (22) $\neg C \in a$ follows. Then by A7′, we have (23) $B \to \neg(B \to B) \in a$ or (24) $C \to \neg(C \to C) \in a$. But 23 contradicts 15 and 24 contradicts 16. Consequently, $B \wedge C \in z$.

Given bi and bii, $z \in K^T$. In addition, we have:

(biii) $b \subseteq z$ and $d \subseteq z$: We prove that b is included in z (the proof for d is similar). Let (25) $B \in b$ but (26) $B \notin z$, i.e., (27) $B \to \neg C \in a$, C being a theorem of L $\cup\{t58\}$. By 1, 25, and 27, we have (28) $\neg C \in c$, contradicting the w-consistency of c.

Consider now the set $u = \{B \mid \exists C[C \to B \in a \ \& \ C \in z]\}$. By Lemma 1.24, we have $u \in K^T$ and $R^T azu$. Next, u is proved w-consistent.

(biv) u is w-consistent: Suppose (29) $\neg B \in u$, B being a theorem of L $\cup\{t58\}$. Then, for some $C \in \mathcal{F}$, (30) $C \to \neg B \in a$ and (31) $C \in z$ follow. But by 31 and definition of z, we have (32) $C \to \neg B \notin a$, contradicting 30.

In sum, we have $z \in K^T$ and $u \in S^T$ such that $R^T azu$, $b \subseteq u$, and $d \subseteq u$. It remains to extend z and u to the required $x \in K^P$ and $y \in S^P$ such that $R^P axy$, $b \leq^P x$ and $d \leq^P x$ (remark that by Corollary 1.23, for any $a, b \in K^P$, $a \leq^P b$ iff $a \subseteq b$), by using Lemmas 2.13 and 1.19.

(c) pt59, $(Rabc \ \& \ Rade \ \& \ a \in O \ \& \ c \in S \ \& \ e \in S) \Rightarrow \exists x \in S \ Rdbx$, is provable in M_C: Let $b, d \in K^P$, $a \in O^P$, $c, e \in S^P$ and (1) $R^P abc$, (2) $R^P ade$. Define the L $\cup\{t59\}$-theory $y = \{B \mid \exists A[A \to B \in d \ \& \ A \in b]\}$ such that (3) $R^T dby$. We prove that y is w-consistent. Suppose, for reductio, (4) $\neg A \in y$, A being a theorem of L $\cup\{t59\}$. By definition of y, we have, for some $B \in \mathcal{F}$, (5) $B \to \neg A \in d$ and (6) $B \in b$. By Red and 5, we get (7) $\neg B \in d$. Given that $a \in O^P$, by t59, either (8) $\neg\neg B \in a$ or (9) $\neg B \in a$ follows. Suppose 8. By A7′, (10) $\neg\neg B \to [\neg B \to \neg(\neg B \to \neg B)]$ is provable in L $\cup\{t59\}$. By 8 and 10, we have (11) $\neg B \to \neg(\neg B \to \neg B) \in a$ and by 2, 7, and 11, (12) $\neg(\neg B \to \neg B) \in e$ follows, contradicting the w-consistency of e. Then, suppose 9. By A7′, we have (13) $\neg B \to [B \to \neg(B \to B)]$, and by 9 and 13, (14) $B \to \neg(B \to B) \in a$, whence, by 1 and 6, we get (15) $\neg(B \to B) \in c$, the consistency of c being contradicted. Consequently, y is w-consistent. Finally y is extended to some $x \in S^P$ such that $R^P dbx$, as it was required.

(d) pt64, $(Rabc \ \& \ Rade \ \& \ a \in O \ \& \ c \in S) \Rightarrow (d \leq a \text{ or } b \leq e)$, is provable in M_C: Let $b, d, e \in K^P$, $a \in O^P$, $c \in S^P$ and (1) $R^P abc$, (2) $R^P ade$. By Corollary 1.23, it suffices to prove $d \subseteq a$ or $b \subseteq e$. Suppose, for reductio that $d \not\subseteq a$ and $b \not\subseteq e$, i.e., that there are $A, B \in \mathcal{F}$ such that (3) $A \in d$, (4) $A \notin a$, (5) $B \in b$, and (6) $B \notin e$. Given 2, 3, and 6, (7) $A \to B \notin a$ follows; and given

$a \in O^P$, by pt64, we have (8) $(A \vee \neg B) \vee (A \to B) \in a$. By 4, 7, and 8, we get (9) $\neg B \in a$. Next, we apply A7', (10) $\neg B \to [B \to \neg(B \to B)]$. Then, we have the following. By 9 and 10, (11) $B \to \neg(B \to B) \in a$; and by 1, 5, and 11, (12) $\neg(B \to B) \in c$. But 12 contradicts the w–consistency of c. Thus, $d \subseteq a$ or $b \subseteq e$ (i.e., $d \leq^P a$ or $b \leq^P e$), as it was to be proved. $\square$

Proposition 6.8 (Pr. of pt41–pt69 in can. EB$_+$-m. c & EB$_{K+}$-m. b). *Let L be an EB$_{cS}$-logic and for any $j(41 \leq j \leq 69)$ let the canonical L $\cup\{tj\}$-model, M_C, be a canonical EB$_+$-model of type c or a canonical EB$_{K+}$-model of type b. Then, ptj is provable in M_C.*

Proof. (I) *pt41–pt64.* pt41–pt64 are proved exactly as in Proposition 6.7 except that each new theory introduced has to be shown non-empty (but recall that if L is an EB$_{K+}$-logic, then any non-empty L-theory is regular; cf. Lemma 3.12). However, this is immediately proved by using Lemma 2.15 (EB$_+$-logics) or Lemma 3.17 (EB$_{K+}$-logics) for any pt41–pt69 save pt58. Let us show it (we use Lemma 2.15; the proofs by using Lemma 3.17 are similar). Consider, for example, pt52 as it is proved in Proposition 6.7. By Lemma 2.15, z is a non-empty L $\cup\{t52\}$-theory since now a and b are supposed to be non-empty and we have $R^P abz$ (and so $R^{NT} abz$). Then, by using the same Lemma 2.15, u is a non-empty L $\cup\{t52\}$-theory as well, since $b, z \in K^{NT}$ and $R^P bzu$ (and so $R^{NT} bzu$). Regarding pt58, the reader is again referred to its proof in Proposition 6.7. Firstly, notice that by Lemma 2.17, $a \in K^{WP}$, since $R^P abc$ and $c \in S^P$. Then, the L $\cup\{t58\}$-theory $z = \{B \mid \forall A \in \mathcal{F}[\vdash_{L \cup\{t58\}} A \Rightarrow (B \to \neg A \notin a)]\}$ is clearly non-empty. Otherwise, $(A \to A) \to \neg A \in a$ (A is an L $\cup\{t58\}$-theorem) whence by Red, $\neg(A \to A) \in a$, contradicting the w-consistency of a. Finally, the set u is also a non-empty member in K^T as $a, z \in K^{NT}$ and $R^P azu$ (and so $R^{NT} azu$).

(II) *pt65–pt69.* Concerning pt65–pt69, we prove pt65, pt67, and pt69 as a way of an example. (pt65 and pt67 are proved for EB$_+$-models; the proof for EB$_{K+}$-models is similar.)

(a) pt65, *(Rabc & $a \in O$ & $c \in S$) $\Rightarrow b \leq a$, is provable in M_C:* Let $a \in O^P$, $b \in K^{NP}$, and $c \in S^{NP}$ such that (1) $R^{NP} abc$, and suppose, for reductio, (2) $b \not\leq^P a$, i.e., (3) $A \in b$ but (4) $A \notin a$ for some $A \in \mathcal{F}$. By t65, (5) $\neg\neg(A \to A) \to (A \vee \neg A)$ is a theorem of L $\cup\{t65\}$, and by Proposition 2.7, we have (6) $\neg\neg(A \to A) \in a$. Thus, (7) $A \vee \neg A \in a$ follows by 5 and 6; and (8) $\neg A \in a$, by 4 and 7. Next, we apply A7 (9) $\neg A \to [A \to \neg(A \to A)]$: by 8 and 9 we get (10) $A \to \neg(A \to A) \in a$ and, finally, by 1, 3, and 10, (11) $\neg(A \to A) \in c$, contradicting the w-consistency of c.

(b) pt67, $(Rabc \ \& \ Rcde \ \& \ a \in O \ \& \ e \in S) \Rightarrow d \leq a$, *is provable in* M_C: Let $b, c, d \in K^{NP}$, $a \in O^P$ and $e \in S^{NP}$ such that (1) $R^{NP}abc$ and (2) $R^{NP}cde$; suppose, for reductio, that there is $A \in \mathcal{F}$ such that (3) $A \in d$ but (4) $A \notin a$. By t67, we have (5) $A \vee (B \to \neg A) \in a$ for some $B \in b$. So, (6) $B \to \neg A \in a$ follows by 4 and 5. By 1 and 6, we have (7) $\neg A \in c$ (since $B \in b$). Next, A7 (8) $\neg A \to [A \to \neg(A \to A)]$ is applied and we get (9) $A \to \neg(A \to A) \in c$ by 7 and 8, whence by 2 and 3, we obtain (10) $\neg(A \to A) \in e$, contradicting the w-consistency of e.

(c) pt69, $(Rabc \ \& \ Rade \ \& \ c \in S) \Rightarrow (a \leq e \ or \ b \leq e)$, *is provable in* M_C: Let $a, b, d, e \in K^{NP}$ and $c \in S^{NP}$ such that (1) $R^{NP}abc$ and (2) $R^{NP}ade$; suppose, for reductio, that there are $A, B \in \mathcal{F}$ such that (3) $A \in a$, (4) $A \notin e$ (5) $B \in b$, and (6) $B \notin e$. We have (7) $A \vee B \in a$ (by 3 and A4); (8) $A \vee B \in b$ (by 5 and A4); (9) $A \vee B \notin e$ (by 4 and 6); (10) $\neg(A \vee B) \notin a$ (by 1 and 8, as $c \in S$), and (11) $C \to (A \vee B) \notin a$ for some $C \in d$ (by 2 and 9). By t69, (12) $(A \vee B) \to [\neg(A \vee B) \vee [C \to (A \vee B)]]$ is a theorem of L $\cup\{t69\}$. By 7 and 12, we have (13) $\neg(A \vee B) \vee [C \to (A \vee B)] \in a$. But 10 and 11 contradict 13. (It is interesting to compare the proof of pt69 just given with that of pt39 in Proposition 6.5.) □

Proposition 6.9 (Pr. of pt41–pt71 in can. EB$_+$-m. d & EB$_{K+}$-m. c). *Let* L *be an* EB$_c$-*logic and for any* $j(41 \leq j \leq 71)$ *let the canonical* L $\cup\{tj\}$-*model,* M_C, *be a canonical* EB$_+$-*model of type d or a canonical* EB$_{K+}$-*model of type c. Then,* ptj *is provable in* M_C.

Proof. (I) *pt41–pt69*. Postulates pt41–pt69 are proved exactly as in Proposition 6.8 except that each new theory introduced has to be shown w-consistent. But this is easy to prove by using Lemma 2.17 (cf. Proposition 6.5). Let us propose an example:

pt55, $(R^2 abcd \ \& \ d \in S) \Rightarrow \exists x, y \in K \ \exists z \in S(Racx \ \& \ Rbcy \ \& \ Rxyz)$, *is provable in* M_C: As $z \in S^P$, the w-consistency of x and y follows immediately by Lemma 2.17.

(II) *pt70 and pt71*. The proof of both postulates leans on the w-consistency of each element in the canonical set K. We prove that pt70 holds in M_C (the proof for pt71 is similar).

pt70, $\exists x \in K \ Raax$, *is provable in* M_C: Define the L $\cup\{t70\}$-theory $y = \{B \mid \exists A[A \to B \in a \ \& \ A \in a]\}$ such that $R^T aay$. If y is w-inconsistent, then there is $B \in \mathcal{F}$ and L $\cup\{t70\}$-theorem A such that (1) $B \to \neg A \in a$ and (2) $B \in a$. By Red and 1 we have (3) $\neg B \in a$ and, finally by t70, 2, and 3, we get (4) $C \in a$, for arbitrary C, contradicting the w-consistency of a.

Then, y is extended to the required prime $L \cup \{t70\}$-theory x such that $R^P aax$. □

From Propositions 6.6–6.9, it follows that any ERB_c-logic can be extended with t41–t64 and that any EB_{cS}-logic can be extended with t41–t69; however, t70 and t71 demand EB_c-logics. More generally (taking into account the list of positive axioms), we have the following.

Theorem 6.10 (Sound. and compl. of EB_+-logics def. from t1–t71). *Let L be an EB_+-logic and L-models be an RM-semantics for L (cf. Definition 1.7). Then, $L \cup \{ptj\}$-models (i.e., L-models with addition of ptj) are an RM-semantics for $L \cup \{tj\}$ (i.e., for the strengthening of L with tj) in the following cases:*
1. *For any $j(1 \leq j \leq 23)$.*
2. *For any $j(1 \leq j \leq 31)$, if L is an EB_{K+}-logic.*
3. *For any $j(1 \leq j \leq 31$ or $41 \leq j \leq 69)$, if L is an EB_{cS}-logic.*
4. *For any $j(1 \leq j \leq 71)$, if L is an EB_c-logic.*
5. *For any $j(1 \leq j \leq 23)$ or $(41 \leq j \leq 64)$, if L is an ERB_c-logic.*

Proof. Given that L is sound and complete w.r.t. L-models, it suffices to prove that (1) tj is true in any $L \cup \{ptj\}$-model; and (2) ptj is provable in the canonical $L \cup \{tj\}$-model. Then, (1)–(5) are proved as follows. (1): by Proposition 6.1 and Proposition 6.2; (2): by Proposition 6.1 and Proposition 6.3; (3): by Propositions 6.1 and 6.6, and Propositions 6.4 and 6.8; (4): by Propositions 6.1 and 6.6, and Propositions 6.5 and 6.9; and finally, (5): by Propositions 6.1 and 6.6, and Propositions 6.2 and 6.7. □

In the following chapter, some particular instances of the systems described in this theorem are briefly commented.

6.3 EXTENSIONS AND EXPANSIONS WITH f-AXIOMS

Consider the following theses formulated in the language of B_{+f} and B_{K+f} (cf. §4.1):

List of f-axioms:

$$t1_f.\ [(A \rightarrow B) \wedge (B \rightarrow f)] \rightarrow (A \rightarrow f)$$
$$= [(A \rightarrow B) \wedge \neg B] \rightarrow \neg A$$
$$t2_f.\ (B \rightarrow f) \rightarrow [(A \rightarrow B) \rightarrow (A \rightarrow f)]$$
$$= \neg B \rightarrow [(A \rightarrow B) \rightarrow \neg A]$$

t3$_f$. $(A \rightarrow B) \rightarrow [(B \rightarrow f) \rightarrow (A \rightarrow f)]$

   $= (A \rightarrow B) \rightarrow (\neg B \rightarrow \neg A)$

t4$_f$. $[A \wedge (A \rightarrow f)] \rightarrow f$

   $= \neg (A \wedge \neg A)$

t5$_f$. $[A \rightarrow (A \rightarrow f)] \rightarrow (A \rightarrow f)$

   $= (A \rightarrow \neg A) \rightarrow \neg A$

t6$_f$. $A \rightarrow [[A \rightarrow (A \rightarrow f)] \rightarrow f]$

   $= A \rightarrow \neg (A \rightarrow \neg A)$

t7$_f$. $[A \rightarrow (B \rightarrow f)] \rightarrow [(A \rightarrow B) \rightarrow (A \rightarrow f)]$

   $= (A \rightarrow \neg B) \rightarrow [(A \rightarrow B) \rightarrow \neg A]$

t8$_f$. $(A \rightarrow B) \rightarrow [[A \rightarrow (B \rightarrow f)] \rightarrow (A \rightarrow f)]$

   $= (A \rightarrow B) \rightarrow [(A \rightarrow \neg B) \rightarrow \neg A]$

t9$_f$. $[A \rightarrow (B \rightarrow f)] \rightarrow [(A \wedge B) \rightarrow f)]$

   $= (A \rightarrow \neg B) \rightarrow \neg (A \wedge B)$

t10$_f$. $[(A \rightarrow A) \wedge (B \rightarrow B)] \rightarrow f] \rightarrow f$

   $= \neg\neg [(A \rightarrow A) \wedge (B \rightarrow B)]$

t11$_f$. $A \Rightarrow (A \rightarrow f) \rightarrow f$

   $= A \Rightarrow \neg\neg A$

t12$_f$. $A \rightarrow [[A \rightarrow (B \rightarrow f)] \rightarrow (B \rightarrow f)]$

   $= A \rightarrow [(A \rightarrow \neg B) \rightarrow \neg B]$

t13$_f$. $[A \rightarrow [B \rightarrow (C \rightarrow f)]] \rightarrow [B \rightarrow [A \rightarrow (C \rightarrow f)]]$

   $= [A \rightarrow (B \rightarrow \neg C] \rightarrow [B \rightarrow (A \rightarrow \neg C)]$

t14$_f$. $A \rightarrow [(A \rightarrow f) \rightarrow f]$

   $= A \rightarrow \neg\neg A$

t15$_f$. $[A \rightarrow (B \rightarrow f)] \rightarrow [B \rightarrow (A \rightarrow f)]$

   $= (A \rightarrow \neg B) \rightarrow (B \rightarrow \neg A)$

t16$_f$. $(A \wedge B) \rightarrow [[A \rightarrow (B \rightarrow f)] \rightarrow f]$

   $= (A \wedge B) \rightarrow \neg (A \rightarrow \neg B)$

t17$_f$. $(A \rightarrow B) \rightarrow [[A \wedge (B \rightarrow f)] \rightarrow f]$

   $= (A \rightarrow B) \rightarrow \neg (A \wedge \neg B)$

t18$_f$. $[A \wedge (B \rightarrow f)] \rightarrow [(A \rightarrow B) \rightarrow f]$

   $= (A \wedge \neg B) \rightarrow \neg (A \rightarrow B)$

t19$_f$. $B \rightarrow [[A \rightarrow (B \rightarrow f)] \rightarrow (A \rightarrow f)]$

$\quad = B \rightarrow [(A \rightarrow \neg B) \rightarrow \neg A]$

t20$_f$. $f \rightarrow (f \rightarrow f)$

$\quad = f \rightarrow \neg f$

t22$_f$. $(A \rightarrow f) \vee (f \rightarrow A)$

$\quad = \neg A \vee (f \rightarrow A)$

t24$_f$. $[(A \wedge B) \rightarrow f] \rightarrow [(A \rightarrow f) \vee (B \rightarrow f)]$

$\quad = \neg(A \wedge B) \rightarrow (\neg A \vee \neg B)$

t25$_f$. $A \rightarrow (f \rightarrow f)$

$\quad = A \rightarrow \neg f$

t26$_f$. $(A \rightarrow f) \rightarrow [B \rightarrow (A \rightarrow f)]$

$\quad = \neg A \rightarrow (B \rightarrow \neg A)$

t27$_f$. $f \rightarrow (A \rightarrow f)$

$\quad = f \rightarrow \neg A$

t28$_f$. $f \rightarrow [B \rightarrow (C \rightarrow f)]$

$\quad = f \rightarrow (B \rightarrow \neg C)$

t29$_f$. $(A \vee f) \rightarrow [(A \rightarrow f) \rightarrow f]$

$\quad = (A \vee f) \rightarrow \neg\neg A$

t32$_f$. $(A \rightarrow B) \vee [(A \rightarrow B) \rightarrow f]$

$\quad = (A \rightarrow B) \vee \neg(A \rightarrow B)$

t33$_f$. $A \vee (A \rightarrow f)$

$\quad = A \vee \neg A$

t34$_f$. $A \vee [A \rightarrow [B \vee (B \rightarrow f)]]$

$\quad = A \vee [A \rightarrow (B \vee \neg B)]$

t35$_f$. $A \rightarrow [B \vee [(A \rightarrow B) \rightarrow f]]$

$\quad = A \rightarrow [B \vee \neg(A \rightarrow B)]$

t36$_f$. $[A \vee (B \rightarrow f)] \vee (A \rightarrow B)$

$\quad = (A \vee \neg B) \vee (A \rightarrow B)$

t37$_f$. $(A \rightarrow f) \vee (B \rightarrow A)$

$\quad = \neg A \vee (B \rightarrow A)$

t38$_f$. $A \vee [B \rightarrow (A \rightarrow f)]$

$\quad = A \vee (B \rightarrow \neg A)$

t39$_f$. $B \rightarrow [(B \rightarrow f) \vee (A \rightarrow B)]$

$\quad = B \rightarrow [\neg B \rightarrow \vee (A \rightarrow B)]$

These theses are instances of the positive axioms t1–t41. Some possible instances are missing since they do not seem significative (e.g., t21$_f$ $A \rightarrow [f \rightarrow (A \vee f)]$?). (We have included a translation of each thesis via the definition $\neg A =_{df} A \rightarrow f$; cf. §4.1.)

Consider now the following semantical postulates corresponding to t1$_f$–t20$_f$, t22$_f$, t24$_f$–t29$_f$, t32$_f$–t39$_f$.

List of f-postulates:

pt1$_f$. $(Rabc$ & $c \in S) \Rightarrow \exists x \in K \ \exists y \in S(Rabx$ & $Raxy)$

pt2$_f$. $(R^2 abcd$ & $d \in S) \Rightarrow \exists x \in K \ \exists y \in S(Rbcx$ & $Raxy)$

pt3$_f$. $(R^2 abcd$ & $d \in S) \Rightarrow \exists x \in K \ \exists y \in S(Racx$ & $Rbxy)$

pt4$_f$. $a \in S \Rightarrow \exists x \in S \ Raax$

pt5$_f$. $(Rabc$ & $c \in S) \Rightarrow \exists x \in S \ R^2 abbx$

pt6$_f$. $(Rabc$ & $c \in S) \Rightarrow \exists x \in S \ R^2 baax$

pt7$_f$. $(R^2 abcd$ & $d \in S) \Rightarrow \exists x, y \in K \ \exists z \in S(Racx$ & $Rbcy$ & $Rxyz)$

pt8$_f$. $(R^2 abcd$ & $d \in S) \Rightarrow \exists x, y \in K \ \exists z \in S(Racx$ & $Rbcy$ & $Ryxz)$

pt9$_f$. $(Rabc$ & $c \in S) \Rightarrow \exists x \in S \ R^2 abbx$

pt10$_f$. $a \in S \Rightarrow \exists x \in Z \ \exists y \in S \ Raxy$

$\quad\quad [Za$ iff for all $b, c \in K, Rabc \Rightarrow \exists x \in O \ Rxbc]$

pt11$_f$. $a \in S \Rightarrow \exists x \in O \ \exists y \in S \ Raxy$

pt12$_f$. $(R^2 abcd$ & $d \in S) \Rightarrow \exists x \in S \ R^2 bacx$

pt13$_f$. $(R^3 abcde$ & $e \in S) \Rightarrow \exists x \in S \ R^3 acbdx$

pt14$_f$. $(Rabc$ & $c \in S) \Rightarrow \exists x \in S \ Rbax$

pt15$_f$. $(R^2 abcd$ & $d \in S) \Rightarrow \exists x \in S \ R^2 acbx$

pt16$_f$. $(Rabc$ & $c \in S) \Rightarrow \exists x \in S \ R^2 baax$

pt17$_f$. $(Rabc$ & $c \in S) \Rightarrow \exists x \in K \ \exists y \in S(Rabx$ & $Rbxy)$

pt18$_f$. $(Rabc$ & $c \in S) \Rightarrow \exists x \in K \ \exists y \in S(Rbax$ & $Raxy)$

pt19$_f$. $(R^2 abcd$ & $d \in S) \Rightarrow \exists x \in S \ R^2 bacx$

pt20$_f$. $(Rabc$ & $c \in S) \Rightarrow (a \in S$ or $b \in S)$

pt22$_f$. ($Rabc$ & $Rade$ & $a \in O$ & $c \in S$) $\Rightarrow$ ($d \in S$ or $b \leq e$)

pt24$_f$. ($Rabc$ & $Rade$ & $c \in S$ & $d \in S$) $\Rightarrow$ $\exists x \in K \, \exists y \in S(Raxy$ &
$\qquad b \leq x$ & $d \leq x$)

pt25$_f$. ($Rabc$ & $c \in S$) $\Rightarrow$ $b \in S$

pt26$_f$. ($R^2 abcd$ & $d \in S$) $\Rightarrow$ $\exists x \in S \, Racx$

pt27$_f$. ($Rabc$ & $c \in S$) $\Rightarrow$ $a \in S$

pt28$_f$. ($R^2 abcd$ & $d \in S$) $\Rightarrow$ $a \in S$

pt29$_f$. ($Rabc$ & $c \in S$) $\Rightarrow$ ($a \in S$ & $\exists x \in S \, Rbax$)

pt32$_f$. ($Rabc$ & $Rade$ & $a \in O$ & $e \in S$) $\Rightarrow$ $Rdbc$

pt33$_f$. ($Rabc$ & $a \in O$ & $c \in S$) $\Rightarrow$ $b \leq a$

pt34$_f$. ($Rabc$ & $Rcde$ & $a \in O$ & $e \in S$) $\Rightarrow$ ($b \leq a$ or $d \leq c$)

pt35$_f$. ($Rabc$ & $c \in S$) $\Rightarrow$ $\exists x \in K(Rbax$ & $x \leq a$)

pt36$_f$. ($Rabc$ & $Rade$ & $a \in O$ & $c \in S$) $\Rightarrow$ ($d \leq a$ or $b \leq e$)

pt37$_f$. ($Rabc$ & $Rade$ & $c \in S$) $\Rightarrow$ $b \leq e$

pt38$_f$. ($Rabc$ & $Rcde$ & $e \in S$) $\Rightarrow$ $d \leq a$

pt39$_f$. ($Rabc$ & $Rade$ & $c \in S$) $\Rightarrow$ ($a \leq e$ or $b \leq e$)

As it has been pointed out above, we are interested here in f-logics (logics in the language of $B_{+,f}$ and $B_{K+,f}$) as long as they can define a negation connective. In the present context, $B_{cS,f}$ seems to be the minimal element in this class. And turning to the list of f-axioms, extensions of $B_{+,f}$ with t11$_f$, t26$_f$, and t27$_f$ include $B_{cS,f}$. Consequently, only extensions of systems equivalent to $B_{+,f}$ plus t11$_f$, t26$_f$, and t27$_f$ or extensions of $B_{cS,f}$ can in principle define a negation connective. Nevertheless, extensions of $B_{+,f}$ and $B_{K+,f}$ by means of the list of f-axioms are not devoid of some interest, and so, this third section of Chapter 6 is dedicated to these extensions. We shall define an RM-semantics for all extensions of $B_{+,f}$ and $B_{K+,f}$ with the list of f-axioms by using the list of corresponding f-postulates. Consequently, the numerals in Propositions 6.11–6.16 and Theorem 6.17 that follow refer to the list of f-axioms or f-postulates, as the case may be, unless otherwise stated.

In the first place, we prove the validity of each f-thesis in any $EB_{+,f}$-model or $EB_{K+,f}$-model in which the corresponding postulate holds.

Proposition 6.11 (Validity of t1$_f$–t39$_f$). *Let a class of $EB_{+,f}$-models or $EB_{K+,f}$-models, $\mathcal{M}$, be defined and $M \in \mathcal{M}$. Then, for any j ($1 \leq j \leq 39$), tj$_f$ is true if ptj holds in M.*

Proof. We prove some selected items.

(a) t4$_f$, $[A \wedge (A \to f)] \to f$, *is true in M*: Suppose there is $A \in \mathcal{F}$ and $a \in K$ in M such that (1) $a \vDash A$, (2) $a \vDash A \to f$ but (3) $a \nvDash f$. By 3, we have (4) $a \in S$. Then pt4$_f$ is applied and we get (5) $Raax$ for some $x \in S$. But by 1, 2, and 5, we have (6) $x \vDash f$, whence we get (7) $x \notin S$, a contradiction.

(b) t18$_f$, $[A \wedge (B \to f)] \to [(A \to B) \to f]$, *is true in M*: Suppose there are $A, B \in \mathcal{F}$ and $a \in K$ in M such that (1) $a \vDash A$, (2) $a \vDash B \to f$ but (3) $a \nvDash (A \to B) \to f$. Then, there are $b, c \in K$ such that (4) $Rabc$, (5) $b \vDash A \to B$, and (6) $c \nvDash f$, i.e., (7) $c \in S$. We apply pt18$_f$ to 4 and 7, obtaining (8) $Rbax$ and (9) $Raxy$ for some $x \in K$ and $y \in S$. Then, we have (10) $x \vDash B$ (by 1, 5, and 8) and (11) $y \vDash f$, i.e., (12) $y \notin S$ (by 2, 9, and 10), a contradiction.

(c) t20$_f$, $f \to (f \to f)$, *is true in M*: Suppose there is $a \in K$ in M such that (1) $a \vDash f$ (i.e., $a \notin S$) but (2) $a \nvDash f \to f$. Then, there are $b, c \in K$ such that (3) $Rabc$, (4) $b \vDash f$ (i.e., $b \notin S$), and (5) $c \nvDash f$ (i.e., $c \in S$). By pt20$_f$, 3, and 5, we have (6) $a \in S$ or (7) $b \in S$. But 6 contradicts 1, and 7 contradicts 4.

(d) t22$_f$, $(A \to f) \vee (f \to A)$, *is true in M*: Suppose there is $A \in \mathcal{F}$ and $a \in O$ in M such that (1) $a \nvDash A \to f$ and (2) $a \nvDash f \to A$. Then, there are $b, c, d, e \in K$ such that (3) $Rabc$, (4) $b \vDash A$, (5) $c \nvDash f$ (i.e., (6) $c \in S$), (7) $Rcde$, (8) $d \vDash f$ (i.e., (9) $d \notin S$), and (10) $e \nvDash A$. By applying pt22$_f$ to 3, 6, and 7 (as $a \in O$), we get (11) $d \in S$ or (12) $b \leq e$. But 11 contradicts 9, and on the other hand, by 4 and 12, (13) $e \vDash A$ is derivable, contradicting 10.

(e) t29$_f$, $(A \vee f) \to [(A \to f) \to f]$, *is true in M*: Suppose there is $A \in \mathcal{F}$ and $a \in K$ in M such that (1) $a \vDash A \vee f$ but (2) $a \nvDash (A \to f) \to f$. Then, there are $b, c \in K$ such that (3) $Rabc$, (4) $b \vDash A \to f$, and (5) $c \nvDash f$ (i.e., (6) $c \in S$). By pt29$_f$, 3, and 6, we have (7) $a \in S$ (i.e., (8) $a \nvDash f$), and (9) $Rbax$ for $x \in S$. Now, by 1 and 8, we get (10) $a \vDash A$. Finally, we have (11) $x \vDash f$ (by 4, 9, and 10). But 11 contradicts 9.

(f) t33$_f$, $A \vee (A \to f)$, *is true in M*: Suppose there is $A \in \mathcal{F}$ and $a \in O$ in M such that (1) $a \nvDash A$ and (2) $a \nvDash A \to f$. Then, there are $b, c \in K$ such that (3) $Rabc$, (4) $b \vDash A$, and (5) $c \nvDash f$ (i.e., (6) $c \in S$). Given $a \in O$, by pt33$_f$, 3, and 6, we have (7) $b \leq a$, whence by 4, (8) $a \vDash A$ follows, contradicting 1.

(g) t35$_f$, $A \to [B \vee [(A \to B) \to f]]$, *is true in M*: Suppose there is $A \in \mathcal{F}$ and $a \in K$ in M such that (1) $a \vDash A$ but (2) $a \nvDash B$ and (3) $a \nvDash (A \to B) \to f$. Then, there are $b, c \in K$ such that (4) $Rabc$, (5) $b \vDash A \to B$, and (6) $c \nvDash f$ (i.e., (7) $c \in S$). By applying pt35$_f$ to 4 and 7, we get (8) $Rbax$ and (9) $x \leq a$ for some $x \in K$. By 1, 5, and 8, we have (10) $x \vDash B$; and by 9 and 10, (11) $a \vDash B$. But 2 and 11 contradict each other.

(h) $t38_f$, $A \vee [B \to (A \to f)]$, *is true in* M: Suppose there are $A, B \in \mathcal{F}$ and $a \in K$ in M such that (1) $a \nvDash A$ but (2) $a \nvDash B \to (A \to f)$. Then, there are $b, c \in K$ such that (3) $Rabc$, (4) $b \vDash B$, and (5) $c \nvDash A \to f$. By 5, there are $d, e \in K$ such that (6) $Rcde$, (7) $d \vDash A$, and (8) $e \nvDash f$ (i.e., (9) $e \in S$). Given 3, 6, and 9, pt38$_f$ is applicable and we have (10) $d \leq a$, whence, by 7, (11) $a \vDash A$ follows, contradicting 1.

Notice that t22$_f$ and t33$_f$ have been proved for EB$_{+,f}$-models. The proof for EB$_{K+,f}$-models is similar: it suffices to delete all references to the set O; more precisely, to change any reference to set O for a corresponding reference to the set K. (The same remark applies to t10$_f$, t11$_f$, t32$_f$–t34$_f$, and t36$_f$.)   □

Next, we proceed to the proof of the adequacy of the postulates. We assume canonical EB$_{+,f}$-models of the form $(K^P, O^P, S^P, R^P, \vDash^P)$ (type (a); there is a type (b) defined below) and canonical EB$_{K+,f}$-models of the form $(K^{NP}, S^{NP}, R^{NP}, \vDash^{NP})$ (type (a); there is a type (b) defined below) (cf. Definition 4.7). On the other hand, we distinguish between postulates pt1$_f$–pt20$_f$, pt22$_f$, pt24$_f$, pt32$_f$–pt36$_f$ provable in any EB$_{+,f}$-model or EB$_{K+,f}$-model and pt25$_f$–pt29$_f$, and pt37$_f$–pt39$_f$ only provable in the latter type of canonical models. (Recall that S^P is now interpreted as the set of all u–consistent prime theories —cf. §4.1—, although in many cases a theory is u–consistent iff it is w–consistent —cf. Proposition 4.28 and the comment following it.)

Proposition 6.12 (Proof of f-postulates in can. EB$_{+,f}$-models). *Let L be an EB$_{+,f}$-logic and for any $j \in \{1–20, 22, 24, 32–36\}$ let the canonical L $\cup\{tj_f\}$-model, M_C, be a canonical EB$_{+,f}$-model of type a. Then, ptj$_f$ is provable in M_C.*

Proof. We consider the postulates pt4$_f$, pt18$_f$, pt20$_f$, pt22$_f$, pt33$_f$, and pt35$_f$ treated in Proposition 6.11 (cf. §4.1).

(a) pt4$_f$, $a \in S \Rightarrow \exists x \in S\ Raax$, *is provable in* M_C: Let (1) $a \in S^P$. According to Lemma 1.24, the set $y = \{B \mid \exists A[A \to B \in a\ \&\ A \in a]\}$ is an L $\cup\{t4_f\}$-theory such that (2) $R^T aay$. Suppose (3) $f \in y$. Then, there is $A \in \mathcal{F}$ such that (4) $A \to f \in a$ and (5) $A \in a$. By t4$_f$, (6) $[A \wedge (A \to f)] \to f$ is an L $\cup\{t4_f\}$-theorem. Then, (7) $f \in a$ follows by 4, 5, and 6. But 7 contradicts 1. Thus, $y \in S^T$. It remains to extend y to some $x \in S^P$ such that $R^P aax$, which is immediate by Lemma 1.17: given $f \notin y$, there is some $x \in K^P$ such that $y \subseteq x$ and $f \notin x$. Clearly, $R^P aax$. Finally, $x \in S^P$ since $f \notin x$.

(b) $pt18_f$, $(Rabc$ & $c \in S) \Rightarrow \exists x \in K \ \exists y \in S(Rabx$ & $Raxy)$, *is provable in* M_C: Let $a, b \in K^P$, $c \in S^P$ and (1) $R^P abc$. Define the L $\cup\{t18_f\}$-theory $z = \{B \mid \exists A[A \rightarrow B \in b \ \& \ A \in a]\}$ such that (2) $R^T baz$. Then, define the L $\cup\{t18_f\}$-theory $u = \{B \mid \exists A[A \rightarrow B \in a \ \& \ A \in z]\}$ such that (3) $R^T azu$ (cf. Lemma 1.24 on the definition of z and u). Suppose (4) $f \in u$. Then, there are $A, B \in \mathcal{F}$ such that (5) $A \rightarrow f \in a$, (6) $B \rightarrow A \in b$, and (7) $B \in a$. By t18$_f$, (8) $[B \wedge (A \rightarrow f)] \rightarrow [(B \rightarrow A) \rightarrow f]$ is an L $\cup\{t18_f\}$-theorem. Then, we have (9) $(B \rightarrow A) \rightarrow f \in a$ by 5, 7, and 8, and finally, (10) $f \in c$ (by 1, 6, and 9), which is impossible, since $c \in S^P$. It remains to extend z and u to the required $x \in K^P$ and $y \in S^P$ such that $R^P abx$ and $R^P axy$, which is done in the customary way.

(c) $pt20_f$, $(Rabc$ & $c \in S) \Rightarrow (a \in S$ or $b \in S)$, *is provable in* M_C: Let $a, b \in K^P$, $c \in S^P$ and (1) $R^P abc$. Suppose, for reductio, (2) $a \notin S^P$ (i.e., $f \in a$) and (3) $b \notin S^P$ (i.e., $f \in b$). By t20$_f$, we have (4) $f \rightarrow (f \rightarrow f)$. Then, we get (5) $f \rightarrow f \in a$ (by 2 and 4) and (6) $f \in c$ (by 1, 3, and 5), contradicting the u-consistency of c ($c \in S^P$).

(d) $pt22_f$, $(Rabc$ & $Rade$ & $a \in O$ & $c \in S) \Rightarrow (d \in S$ or $b \leq e)$, *is provable in* M_C: Let $b, d, e \in K^P$, $a \in O^P$, $c \in S^P$ and (1) $R^P abc$ and (2) $R^P ade$. Suppose, for reductio, (3) $f \in d$ and (4) $A \in b$ but (5) $A \notin e$, for some $A \in \mathcal{F}$. We have (6) $A \rightarrow f \notin a$ by 1 and 4, since $f \notin c$; and (7) $A \rightarrow f \in a$ or $f \rightarrow A \in a$ by t22$_f$ since $a \in O^P$. Thus, by 6 and 7, we get (8) $f \rightarrow A \in a$. Finally, we have (9) $A \in e$ (by 2, 3, and 8), contradicting 5.

(e) $pt33_f$, $(Rabc$ & $a \in O$ & $c \in S) \Rightarrow b \leq a$, *holds in* M_C: Let $b \in K^P$, $a \in O^P$, $c \in S^P$ and (1) $R^P abc$. And suppose, for reductio, (2) $A \in b$ and (3) $A \notin a$ for some $A \in \mathcal{F}$. By t33$_f$, (4) $A \in a$ or (5) $A \rightarrow f \in a$ follows since $a \in O^P$. Thus, we have (6) $A \rightarrow f \in a$, whence, by 1, 2, and 6, (7) $f \in c$ is derived, contradicting the u-consistency of c ($c \in S^P$).

(f) $pt35_f$, $(Rabc$ & $c \in S) \Rightarrow \exists x \in K(Rbax$ & $x \leq a)$, *holds in* M_C: Let $a, b \in K^P$, $c \in S^P$ and (1) $R^P abc$. Define now the L $\cup\{t35\}$-theory y such that (2) $R^T bay$. Then, suppose, for reductio, that there is $A \in \mathcal{F}$ such that (3) $A \in y$ but (4) $A \notin a$. By definition of y, we have (5) $B \rightarrow A \in b$ and (6) $B \in a$ for some $B \in \mathcal{F}$. Next, t35$_f$ is applied in the form (7) $B \rightarrow [A \vee [(B \rightarrow A) \rightarrow f]]$. We have (8) $A \vee [(B \rightarrow A) \rightarrow f] \in a$ (by 6 and 7); (9) $(B \rightarrow A) \rightarrow f \in a$ (by 4 and 8) and, finally, (10) $f \in c$ (by 1, 5, and 9). But 10 contradicts the u-consistency of c ($c \in S^P$). Finally, y is extended to the required prime L $\cup\{t35_f\}$-theory x. $\square$

Corollary 6.13 (Proof of f-postulates in can. EB$_{K+f}$-models I). *Let L be an EB$_{K+f}$-logic and for any $j \in \{1-20, 22, 24, 32-36\}$ let the canonical L*

$\cup \{tj_f\}$-*model*, M_C, *be a canonical* EB_{K+f}-*model of type a. Then*, ptj_f *is provable in* M_C.

Proof. Given Proposition 6.12, it suffices to prove that the new theories introduced in the referred postulates are non-empty. But non-emptiness of new introduced theories either follows from their definition (as it is the case with pt10 and pt11) or else by Lemma 3.17, as it is the case with the rest of the semantical postulates (cf. Proposition 6.3). □

Proposition 6.14 (Proof of f-postulates in can. EB_{K+f}-models II). *Let L be an* EB_{K+f}-*logic and for any* $j \in \{1–20, 22, 24–29, 32–39\}$ *let the canonical L* $\cup \{tj_f\}$-*model*, M_C, *be a canonical* EB_{K+f}-*model of type a. Then*, ptj_f *is provable in* M_C.

Proof. That $pt1_f$–$pt20_f$, $pt22_f$, $pt24_f$, $pt32_f$–$pt36_f$ are provable was shown in Corollary 6.13. Concerning the rest of the postulates, a couple of examples will be sufficient. (We prove $pt29_f$ and $pt38_f$ used in Proposition 6.11.)

(f) $pt29_f$, $(Rabc \;\&\; c \in S) \Rightarrow (a \in S \;\&\; \exists x \in S \; Rbax)$, holds in M_C: (The following theorems ta_f and tb_f are derivable in $B_{+f} \cup \{t29_f\}$ and will be used in the subsequent proof: $ta_f \; A \to [(A \to f) \to f]$; $tb_f \; f \to (A \to f)$.) Let $a, b \in K^{NP}$, $c \in S^{NP}$ and (1) $R^{NP}abc$.

(i) $a \in S^{NP}$: Suppose (2) $a \notin S^{NP}$ (i.e., $f \in a$) and let (3) $A \in b$ (theories are non-empty in M_C). By 2 and tb_f, we get (4) $A \to f \in a$, whence (5) $f \in c$ follows (by 1, 3, and 4), contradicting the u-consistency of c ($c \in S^{NP}$). Consequently, $a \in S^{NP}$.

(ii) $\exists x \in S^{NP} \; R^{NP}bax$: Define the L $\cup\{t29_f\}$theory x such that $R^{NP}bax$. Suppose (6) $x \notin S^{NT}$ (i.e., $f \in x$). Then, there is $A \in \mathcal{F}$ such that (7) $A \to f \in b$ and (8) $A \in a$. By ta_f and 8, we have (9) $(A \to f) \to f \in a$, whence (10) $f \in c$ follows by 7 and 9, since $R^{NP}abc$, by 1. But 10 is impossible. Consequently, $x \in S^{NT}$.

It remains to extend x to the required u-consistent prime L $\cup\{t29_f\}$-theory, which is done in the customary way.

(g) $pt38_f$, $(Rabc \;\&\; Rcde \;\&\; e \in S) \Rightarrow d \le a$, is provable in M_C: Let $a, b, c, d \in K^{NP}$, $e \in S^{NP}$, (1) $R^{NP}abc$ and (2) $R^{NP}cde$. For reductio, suppose that there are $A, B \in \mathcal{F}$ such that (3) $A \in d$ and (4) $A \notin a$ and let (5) $B \in b$ (theories are non-empty). By $t38_f$, we have (6) $A \vee [B \to (A \to f)] \in a$. Thus, (7) $B \to (A \to f) \in a$ follows by 4 and 6. Then, we get (8) $A \to f \in c$ (by 1, 5, and 7) and, finally, (9) $f \in e$ (by 2, 3, and 8). But 9 contradicts the u-consistency of e ($e \in S^{NP}$). □

So far in this section, we have taken into account only classical propositional tautologies and f-instances thereof in order to define f-systems with intuitionistic type negations. But in the sequel, it is our intention to briefly discuss extensions of EB_{+f}-logics and EB_{K+f}-logics by using the thesis $t0_f\ f \to A$, which is not, of course, an f-instance of a classical tautology, which is validated only if f is false in every $a \in K$ in the models (cf. Definitions 1.4 and 3.3), and, finally, which consequently requires that all theories be u-consistent (cf. Definition 1.15 and §4.1) from the canonical point of view. This last condition means that we need canonical EB_{+f}-models of the form $(K^{\mathrm{UNP}}, O^U,\ R^{\mathrm{UNP}}, \vDash^{\mathrm{UNP}})$ (canonical EB_{+f}-models of type (b)) and canonical EB_{K+f}-models of the form $(K^{\mathrm{UNP}}, R^{\mathrm{UNP}}, \vDash^{\mathrm{UNP}})$ (canonical EB_{K+f}-models of type (b)), where the superscripts N, R, and P are read exactly as in the case of canonical EB_+-models and EB_{K+}-models (cf. Definitions 1.27 and 3.18) and the superscript U abbreviates u-consistent. In order to handle these types of canonical models, we need the lemma proved below. This lemma plays here the role Lemma 2.17 plays in canonical models where theories have to be w-consistent, i.e., canonical EB_+-models of type d and canonical EB_{K+}-models of type c (Definitions 1.27 and 3.18). (Notice that $t0_f$ has been employed in Chapter 4 (there, it was labelled $A8_f$) for axiomatizing logics definitionally equivalent to B_c and B_K. Here $t0_f$ will be used in a more general way.)

Lemma 6.15 (R^P and u-consistency). *Let L be an EB_{+f}-logic or an EB_{K+f}-logic where $t25_f$ and $t27_f$ hold; and let a, b be L-theories and c be an u-consistent L-theory such that $R^T abc$. Then, b and a are u-consistent as well.*

Proof. Let $a, b \in K^T$, $c \in K^{UT}$ and (1) $R^T abc$. (i) Suppose (2) $f \in a$ and let (3) $A \in b$. By $t27_f$, (4) $f \to (A \to f)$ is a theorem, whence by 2, we get (5) $A \to f \in a$ and finally, (6) $f \in c$ (by 1, 3, and 5), contradicting the u-consistency of c. (ii) Suppose (7) $f \in b$ and let (8) $B \in a$. By $t25_f$, (9) $B \to (f \to f)$ is a theorem. Then, by 8 and 9, we have (10) $f \to f \in a$, whence by 1 and 7, (11) $f \in c$ follows, contradicting again the u-consistency of c. Consequently, a and b are u-consistent. $\square$

The proof of this lemma seems to demand the use of $t25_f$ and $t27_f$ and, in general, the use of non-empty theories (cf. lines 3 and 8), which is the reason why canonical EB_{+f}-models of type b and canonical EB_{K+f}-models of type b have been defined as shown above. On the other hand, the condition that all theories be non-empty in the canonical model entails, as discussed above, the presence of a theorem of the form $A \to (B \to C)$

($A \neq B$) in the logic. Therefore, in what follows, we will only investigate B_{+f}-extensions with $t0_f$ $f \to A$ that contain t25 $B \to (A \to A)$ among their theorems (there are several other options that we leave to the reader; cf. §2.6). Then, let L be the result of extending B_{+f} with t25 and $t27_f$ (notice that $t25_f$ is immediate). Firstly, we prove that L can be extended with the theses in the f-list augmented with $t0_f$. Then, we prove a theorem collecting all the results in the section.

Proposition 6.16 (Ext. of $B_{+f} \cup \{t25, t27_f\}$ with the f-axioms list). *Let L_f be an EL-logic and for any $j \in \{0$–$20, 22, 24$–$29, 32$–$39\}$, let the canonical $L_f \cup \{tj\}$-model, M_C, be a canonical EB_{+f}-model of type b or a canonical EB_{K+f}-model of type b. Then ptj is provable in M_C.*

Proof. It is similar to that of Proposition 6.9 except that each new theory introduced has to be shown u-consistent, which is easy by leaning on Lemma 6.15 (cf. the use of Lemma 2.17 for showing the w-consistency of new theories in the proof of Proposition 6.9). (Notice that all references to the set S in the theses in the list of f-axioms are omitted and that the condition $a \in O$ has to be added in $pt37_f$ and $pt38_f$.)   □

The chapter is ended with the following theorem and remark.

Theorem 6.17 (Soundness and completeness of EB_{+f}-logics). *Let L be an EB_{+f}-logic and L-models be an RM-semantics for L. Then, $L \cup \{ptj_f\}$-models are an RM-semantics for $L \cup \{tj_f\}$ in the following cases:*

1. *For any $j \in \{1$–$20, 22, 24, 32$–$36\}$.*
2. *For any $j \in \{1$–$20, 22, 24$–$29, 32$–$39\}$, if L is an EB_{K+f}-logic.*
3. *For any $j \in \{0$–$20, 22, 24$–$29, 32$–$39\}$, if L is an extension of $B_{+f} \cup \{t25(B \to (A \to A))$ and $t27_f\}$.*

Proof. (The numerals in 1, 2, and 3 refer to the theses on the f-axioms list; $t0_f$ is $f \to A$.) The proof is similar to that of Theorem 6.10 by using now the following propositions. (1): Proposition 6.11 and Proposition 6.12; (2): Proposition 6.11 and Proposition 6.14; (3): Proposition 6.11 and Proposition 6.16.   □

In addition, we note the following facts regarding extensions of EB_{+f}-logics by using the positive axioms list (cf. Theorem 6.10):

1. EB_{+f}-logics can be extended with t1–t23.
2. EB_{K+f}-logics can be extended with t1–t31.
3. Logics containing $B_{+f} \cup \{t25 (B \to (A \to A))$ and $t27_f\}$ can be extended with t1–t40.

Remark 6.18 (Strong soundness and completeness). We recall that strong soundness and completeness (of sorts) can be proved for all the systems introduced in the present work (cf. §1.4 and Theorems 1.34 and 3.39).

CHAPTER 7

On some extensions and expansions of the basic logics

7.1 SOME SYSTEMS DEFINABLE FROM t1–t71

Intuitionistic logic, H, Minimal Intuitionistic Logic, H_M, and Intermediate Logics (i.e., logics between Intuitionistic Logic and Classical Logic), ILs, are the best-known logics endowed with intuitionistic-type negations. But, in addition to these logics, there are a number of systems with intuitionistic-type negations definable by t1–t71, some of which are going to concisely be treated in this chapter. Some of these systems are included in H or in a given IL, but others are not. There are even logics with the "Principle of Excluded Middle", $A \vee \neg A$, among its theorems.

Firstly, a series of well-known positive logics is defined. These logics will be extended with intuitionistic-type negations formulable with t41–t71. Then, a few features of the resulting systems will be briefly discussed.

Leaning on t1–t40, a number of interesting positive logics can be defined, some of which are axiomatized below by adding some of t1–t40 to B_+ (these axiomatizations have not necessarily independent axioms and/or rules):

- Ticket Entailment, T_+: t2, t3, t5.
- Entailment Logic, E_+: T_+ plus t10.
- Relevance Logic, R_+: T_+ plus t14.
- 3-valued extension of R-Mingle, $RM3_+$: R_+ plus t20, t33.
- Modal Logic, $S4_+$: T_+ plus t26.
- Modal Logic, $S5_+$: $S4_+$ plus t32.
- Intuitionistic Logic, H_+: R_+ plus t27.
- Dummett's Logic, LC_+: H_+ plus t22.

The relations these logics maintain to each other can be summarized in the following diagram (the arrow stands for set inclusion and C_+ is Positive Classical Logic axiomatizable by adding t33 to H_+):

Routley-Meyer Ternary Relational Semantics for Intuitionistic-Type Negations.
DOI: http://dx.doi.org/10.1016/B978-0-08-100751-8.00009-2

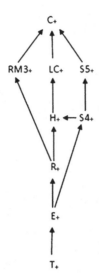

The following, among t1–t40, are derivable in each one of these logics:

- T_+: t1–t5, t7, t8, t9, t17.
- E_+: t1–t5, t7–t11, t17.
- R_+: t1–t19.
- $RM3_+$: t1–t24, t32–t36, t38.
- $S4_+$: t1–t5, t7–t11, t17, t20, t21, t25, t26.
- $S5_+$. t1–t5, t7–t11, t17, t20, t21, t25, t26, t32, t35.
- H_+: t1–t21, t25–t31, t39.
- LC_+: t1–t31, t39.

The full logics H and LC are axiomatized when adding t44 and t70 to H_+ and LC_+, respectively. On the other hand, the full logics T, E, R, RM3, S4, and S5 are formulated by adding to the respective positive system t44, t50, and the strong contraposition axiom $(\neg A \to B) \to (\neg B \to A)$, which does not appear among t1–t71 (concerning these logics the reader can consult [2] and [37] about T, E, R, and RM3; [18] about S4 and S5; and, finally, the appropriate parts, referred to above, of [16] and [44]).

The following, among t41–t71, are derivable in each one of the full logics:

- T: t41, t43, t44, t48–t62.
- E: t41, t43, t44, t48–t62.
- R: t41–t62.
- RM3: t41–t64, t66, t67, t69.
- S4: t41, t43, t44, t48–t63, t65, t68, t70.
- S5: t41, t43, t44, t48–t63, t65, t68, t70.

- H: t41–t57, t68–t71.
- LC: t41–t59, t68–t71.

Notice that the full logics are related as follows (C is classical propositional logic):

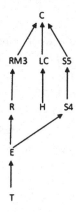

Of course, there are many other logics definable from t1–t71. For instance, intermediate logics such as Jankovic's K_c (H plus t59) or the logic Bd2 (H plus t63) (cf. [29] and the references therein about these logics). Or, to take another example of definable logics by using t1–t71 and the strong contraposition axiom remarked above, we have the interesting contraction-less versions of the logics we have axiomatized, such as TW, EW and RW, which are the result of dropping t5 (positive fragments) and t5 and t50 (full logics) from the formulation of T, E, and R, and their positive fragments, displayed above.

Returning to intuitionistic-type negations, a first question to be remarked is the following. Let L_+ be any of the positive systems above or one of the subsystems thereof formulated with t1–t40; and let L be the result of adding to L_+ axiom A8 and rules Red and VeqDn of B_c together with any selection of t41–t71 (axiom A7 of B_{cS} suffices if the selection is restricted to t41–t69). It follows from Theorem 6.10 that there is an RM-semantics w.r.t. which L is sound and complete. Consider, for example, the logic R_+ strengthened with t41–t64, t67, t68, t70, and t71. An RM-semantics for this system is given by B_+-models supplemented with the semantical postulates P3, P5 (or P4), pt3, pt5, pt14, pt41–pt64, pt67, pt68, pt70, and pt71 together with the appropriate notion of validity (it is obvious that both the system and its semantics can be simplified —there are non-independent theses and postulates).

7.2 NO COLLAPSE BETWEEN CERTAIN SYSTEMS

In the sequel, we comment some aspects of the intuitionistic-type negation expansions of some of the positive logics just formulated. A first interesting point to be remarked is that t41–t69 can be derived in EB_{cS}-logics with certain positive axioms.

Proposition 7.1 (Deriving t41–t69 with some positive theses). *Consider the following pair of theses:* $\langle t1, t41 \rangle$, $\langle t2, t42 \rangle$, $\langle t3, t43 \rangle$, $\langle t15, t44 \rangle$, $\langle t19, t45 \rangle$, $\langle t12, t46 \rangle$, $\langle t13, t47 \rangle$, $\langle t14, t48 \rangle$, $\langle t4, t49 \rangle$, $\langle t5, t50 \rangle$, $\langle t6, t51 \rangle$, $\langle t17, t52 \rangle$, $\langle t18, t53 \rangle$, $\langle t9, t54 \rangle$, $\langle t16, t55 \rangle$, $\langle t7, t56 \rangle$, $\langle t8, t57 \rangle$, $\langle t24, t58 \rangle$, $\langle t32, t59 \rangle$, $\langle t32, t60 \rangle$, $\langle t33, t61 \rangle$, $\langle t35, t62 \rangle$, $\langle t34, t63 \rangle$, $\langle t36, t64 \rangle$, $\langle t40, t65 \rangle$, $\langle t37, t66 \rangle$, $\langle t38, t67 \rangle$, $\langle t26, t68 \rangle$, *and* $\langle t39, t69 \rangle$. *Let L be any* EB_{cS}-logic and $\langle ti, tj \rangle$ anyone of the pairs just remarked. Then, tj is a theorem of L $\cup \{ti\}$.

Proof. The proof is very easy and two or three examples will suffice. Axiom A7 of B_{cS} in the weak form A7' $\neg A \rightarrow [A \rightarrow \neg (A \rightarrow A)]$ and T1 of the same system (i.e., $[B \rightarrow \neg (A \rightarrow A)] \rightarrow \neg B$) will be sufficient for proving all cases.

(a) $\langle t15, t44 \rangle$. By t15, we have (1) $[A \rightarrow [B \rightarrow \neg (B \rightarrow B)]] \rightarrow [B \rightarrow [A \rightarrow \neg (B \rightarrow B)]]$. By A7' and the rule Pref, we get (2) $(A \rightarrow \neg B) \rightarrow [A \rightarrow [B \rightarrow \neg (B \rightarrow B)]]$; and by T1 and the same rule, (3) $[B \rightarrow [A \rightarrow \neg (B \rightarrow \neg B)]] \rightarrow (B \rightarrow \neg A)$. Finally, by 1, 2, 3, and the rule Trans, (4) $(A \rightarrow \neg B) \rightarrow (B \rightarrow \neg A)$, i.e., t44, is derivable.

(b) $\langle t24, t58 \rangle$. We start by t24 in the form (1) $[(A \wedge B) \rightarrow \neg [(A \wedge B) \rightarrow (A \wedge B)]] \rightarrow [[A \rightarrow \neg [(A \wedge B) \rightarrow (A \wedge B)]] \vee [B \rightarrow \neg [(A \wedge B) \rightarrow (A \wedge B)]]]$. Then, we use T1 in the forms (2) $[A \rightarrow \neg [(A \wedge B) \rightarrow (A \wedge B)]] \rightarrow \neg A$ and (3) $[B \rightarrow \neg [(A \wedge B) \rightarrow (A \wedge B)]] \rightarrow \neg B$ and A7' in the form (4) $\neg (A \wedge B) \rightarrow [(A \wedge B) \rightarrow \neg [(A \wedge B) \rightarrow (A \wedge B)]]$. Finally, (5) $\neg (A \wedge B) \rightarrow (\neg A \vee \neg B)$, i.e., t58 is derivable by 1, 2, 3, 4, and B_+.

(c) $\langle t32, t59 \rangle$. By t32, we have (1) $[A \rightarrow \neg (A \rightarrow A)] \vee [[A \rightarrow \neg (A \rightarrow A)] \rightarrow \neg (A \rightarrow A)]$; and by A7' and T1, (2) $\neg A \leftrightarrow [A \rightarrow \neg (A \rightarrow A)]$. Then, (3) $\neg A \vee [\neg A \rightarrow \neg (A \rightarrow A)]$ follows by 1 and 2. Finally, (4) $\neg A \vee \neg \neg A$, i.e., t59, is derivable by 3, T1, $[\neg A \rightarrow \neg (A \rightarrow A)] \rightarrow \neg \neg A$, and B_+.   □

We note a couple of remarks on Proposition 7.1.

Remark 7.2 (The minimal system for Proposition 7.1). As A7 is sufficient in the form $\neg A \rightarrow [A \rightarrow \neg (A \rightarrow A)]$, it is to be remarked that Proposition 7.1 is provable in any extension of B_+ containing T1 of B_{cS} and A7 in the restricted form just indicated. Thus, in particular, Proposition 7.1 is provable in any extension of the constructive relevance logic $\mathcal{R}B_c$.

Remark 7.3 (Non-definability of t70 and t71). Notice that t70 and t71 are not definable via $\neg A =_{df} A \to f$ from a positive thesis: $[A \wedge (A \to C)] \to B$ and $A \to [(A \to C) \to B]$ are not classical tautologies.

However, the reverse road to the one developed in Proposition 7.1 is not in general possible. Let us propose three particular examples.

Consider the following sets of truth-tables STI, STII, and STIII (3 is the only designated value). The tables for $\to$ are:

STI $\to$	0	1	2	3		STII $\to$	0	1	2	3
0	3	3	3	3		0	3	3	3	3
1	0	3	3	3		1	0	3	3	3
2	0	2	3	3		2	0	0	3	3
3	0	1	2	3		3	0	0	2	3

STIII $\to$	0	1	2	3
0	3	3	3	3
1	0	3	3	3
2	0	1	3	3
3	0	0	1	3

And the tables for $\wedge$ and $\vee$ are:

$\wedge$	0	1	2	3		$\vee$	0	1	2	3
0	0	0	0	0		0	0	1	2	3
1	0	1	1	1		1	1	1	2	3
2	0	1	2	2		2	2	2	2	3
3	0	1	2	3		3	3	3	3	3

STI, STII, and STIII verify the axioms and rules of B_+ and, in addition, we have the following proposition (in case a tester is needed for checking this and the propositions to follow, the reader can use that in [17]):

Proposition 7.4 (The reverse road to Proposition 7.1 I). *STI, STII, and STIII verify the following theses in the set t1–t40 (theses omitted are falsified):*
1. *STI: t2, t3, t10–t15, t19–t31, t39.*
2. *STII: t1–t5, t7–t13, t17, t20–t26.*
3. *STIII: t4, t5, t9–t11, t17, t20–t25.*

Proof. It is left to the reader. □

Consider now the following tables for ¬:

	¬			¬
(a) 0	3	(b) 0	3
1	0	1	3
2	0	2	3
3	0	3	3

We have the following proposition (STIa means that the table a for ¬ is added to the tables for →, ∧, and ∨ in STI; STIb, STIIa, etc. are read similarly).

Proposition 7.5 (The reverse road to Proposition 7.1 II). *The following theses in the set t41–t71 are verified by addition of negation tables (a) and (b) to STI, STII, and STIII (theses omitted are falsified):*
1. *STIb, STIIb, and STIIIb: t41–t69.*
2. *STIa: t41–t59, t68–t71.*
3. *STIIa and STIIIa: t41–t59, t68, t70, t71.*

Proof. It is left to the reader. □

From Propositions 7.4 and 7.5, it follows that certain groups of important positive theses are not provable in strong systems with intuitionistic-type negations. In particular, we have:
1. The Axiom of Contraction t5, $[A \to (A \to B)] \to (A \to B)$, and related theses such as t1, t4–t9, and t16–t18 are not derivable in a system equivalent to or included in the one axiomatized by t2, t3, t10–t15, t19–t31, and t39 together with either t41–t69 or t41–t59, t68–t71 (STI).
2. The Axiom of Assertion t14, $A \to [(A \to B) \to B]$, and related theses such as t6, t15, and t19; and, on the other hand, the Axiom Veq t27, $A \to (B \to A)$, and related theses such as t28–t31, t39; and, finally, the Disjunctive Peirce's Law t33, $A \lor (A \to B)$, and related theses such as t32, t34–t38, and t40 are not derivable in any system equivalent to or included in the one axiomatized by t1–t5, t7–t13, t17, t20–t26 together with either t41–t69 or t41–t59, t68, t70, and t71 (STII).
3. The Axiom of Transitivity t3, $(A \to B) \to [(B \to C) \to (A \to C)]$, and related theses such as t1, t2, t7, and t8 are not derivable in any system equivalent to or included in the one axiomatized by t4, t5, t9–t11, t17, t20–t25 together with either t41–t69 or t41–t59, t68, t70, and t71 (STIII).

Related to the question just discussed is the fact that positive systems do not generally collapse when the intuitionistic-type negations here defined are added. Let us propose some examples. Consider the following sets of truth-tables STI, STII, and STIII (1, 2, and 3 are designated values). The tables for $\wedge$ and $\vee$ are the same as the ones used in Propositions 7.4 and 7.5 and the tables for $\rightarrow$ are:

	$\rightarrow$	0	1	2	3
	0	3	3	3	3
STI	1	0	1	3	3
	2	0	0	1	3
	3	0	0	0	3

	$\rightarrow$	0	1	2	3
	0	3	3	3	3
STII	1	0	1	1	3
	2	0	0	1	3
	3	0	0	0	3

	$\rightarrow$	0	1	2	3
	0	3	3	3	3
STIII	1	0	1	2	3
	2	0	0	1	3
	3	0	0	0	3

We have:

Proposition 7.6 (No collapse between positive systems I).
1. *STI verifies the axioms and rules of* T_+ *but falsifies the characteristic axiom of* E_+, *t10,* $[[(A \rightarrow A) \wedge (B \rightarrow B)] \rightarrow C] \rightarrow C$ $(A = B = 1, C = 2)$.
2. *STII verifies the axioms and rules of* E_+ *but falsifies the characteristic axiom of* R_+, *t11,* $A \rightarrow [(A \rightarrow B) \rightarrow B]$ $(A = B = 2)$.
3. *STIII verifies the axioms and rules of* R_+ *but falsifies the characteristic axiom of RM3, t20,* $A \rightarrow (A \rightarrow A)$ $(A = 2)$. *(So, STIII falsifies the characteristic axioms of* $S4_+$ *and* H_+, *t25 and t27, respectively.)*

Proof. It is left to the reader.   □

Consider now the tables for negation:

		$\neg$
	0	3
(a)	1	0
	2	0
	3	0

		$\neg$
	0	3
(b)	1	3
	2	3
	3	3

We have:

Proposition 7.7 (No collapse between positive systems II).
1. *STIa, STIIa, and STIIIa verify t41–t64, t67, t68, t70, and t71.*
2. *STIb, STIIb, and STIIIb verify t41–t69.*

Proof. It is left to the reader.   □

Thus, R_+, for instance, does not collapse in RMO_+, $S4_+$, or H_+ when either t41–t69 or t41–t64, t67, t68, t70, and t71 are added. And we remark that similar results would obtain with the relations holding between the rest of the positive logics considered above.

7.3 RELEVANCE AND PARACONSISTENT LOGICS

The chapter is ended with a couple of notes on relevance logics, paraconsistent logics and intuitionistic-type negations (we have no space for a more detailed treatment).

Consider the following set of truth-tables $SRM3_+$ (1 and 2 are designated values):

$\rightarrow$	0	1	2		$\wedge$	0	1	2		$\vee$	0	1	2
0	2	2	2		0	0	0	0		0	0	1	2
1	0	1	2		1	0	1	1		1	1	1	2
2	0	0	2		2	0	1	2		2	2	2	2

These tables characterize the positive fragment of the logic RM3, which, as it is known, has the variable-sharing property, vsp; for suppose that in $A \rightarrow B$ there are no propositional variables in common to A and B. Assign then 2 to all variables in A and 1 to all variables in B. The value of $A \rightarrow B$ is then 0, whence $A \rightarrow B$ is not a theorem of $RM3_+$. Consider now the following tables for negation:

(a)

	$\neg$
0	2
1	0
2	0

(b)

	$\neg$
0	2
1	2
2	2

We have:

Proposition 7.8 (Constructive negation in $RM3_+$).
1. *$RM3_+a$ verify t41–t64, t67, t68, t70, t71.*
2. *$RM3_+b$ verify t41–t69.*

Proof. It is left to the reader.   □

Therefore, the intuitionistic-type negation characterized by table (a) or that characterized by table (b) can be added to $RM3_+$ without the vsp being violated in the positive fragment. By the way, we remark that neither of the systems characterized by table (a) and table (b) is included in the logic RM3: for instance, t68 is not a theorem of RM3.

On the other hand, the Crystal lattice truth-tables, CL-tables, are $(1, 2, 3, 4,$ and 5 are designated values):

→	0	1	2	3	4	5		¬		∧	0	1	2	3	4	5
0	5	5	5	5	5	5		5		0	0	0	0	0	0	0
1	0	1	2	3	4	5		4		1	0	1	1	1	1	1
2	0	0	2	0	2	5		2		2	0	1	2	1	2	2
3	0	0	0	3	3	5		3		3	0	1	1	3	3	3
4	0	0	0	0	1	5		1		4	0	1	2	3	4	4
5	0	0	0	0	0	5		0		5	0	1	2	3	4	5

∨	0	1	2	3	4	5
0	0	1	2	3	4	5
1	1	1	2	3	4	5
2	2	2	2	4	4	5
3	3	3	4	3	4	5
4	4	4	4	4	4	5
5	5	5	5	5	5	5

We have:

Proposition 7.9 (Some theses of CL-logic). *The CL-tables verify t1–t19, t32–t36, t38; t41–t64, t66, and t67.*

Proof. It is left to the reader.   □

Now, any system verified by the CL-tables has the vsp (cf. [9]). Consequently, any system axiomatized by any selection of the theses remarked in Proposition 7.9 has the vsp. In particular, any extension of RB_c by using any selection of the said theses has the vsp. (Notice that t32–t36, t63, t64, t66, and t67 are not provable in relevance logic R.)

On the other hand, we recall that L is a *paraconsistent logic* if the Rule Ecq $(A \wedge \neg A \Rightarrow B)$ does not hold in L (cf. [11] and references therein). We have:

Proposition 7.10 (Paraconsistent strengthenings). *Let ST be any of the sets of truth-tables in Propositions 7.4–7.8, negation being interpreted according to table (b). And let L be a logic (cf. Definition 1.2) verified by ST. Then L is a paraconsistent logic.*

Proof. Immediate. Let p_i, p_j ($i \neq j$) be propositional variables. Assign p_i the value 3 (2 in the case of the 3-element tables in Proposition 7.8) and p_j the value 0. Then $p_i \wedge \neg p_i$ is assigned a designated value and consequently the rule Ecq is falsified. (Notice that L is paraconsistent in the same sense as Minimal Intuitionistic Logic is paraconsistent. Of course, paraconsistent logics in a stricter sense of the term paraconsistent can be defined from t1–t71 (cf. [28] and references therein).)   □

We remark that a wealth of strong paraconsistent logics have been defined above. For example, expansions with t41–t69 of the positive fragment of RM3 or the positive fragment of S5 (cf. [18]; STIIb in Proposition 7.4).

APPENDIX A

List of axioms and postulates

A.1 POSITIVE AND NEGATION AXIOMS AND THEIR CORRESPONDING POSTULATES

Axioms:

t1. $[(A \to B) \wedge (B \to C)] \to (A \to C)$

t2. $(B \to C) \to [(A \to B) \to (A \to C)]$

t3. $(A \to B) \to [(B \to C) \to (A \to C)]$

t4. $[A \wedge (A \to B)] \to B$

t5. $[A \to (A \to B)] \to (A \to B)$

t6. $A \to [[A \to (A \to B)] \to B]$

t7. $[A \to (B \to C)] \to [(A \to B) \to (A \to C)]$

t8. $(A \to B) \to [[A \to (B \to C)] \to (A \to C)]$

t9. $[A \to (B \to C)] \to [(A \wedge B) \to C]$

t10. $[[(A \to A) \wedge (B \to B)] \to C] \to C$

t11. $A \Rightarrow (A \to B) \to B$

t12. $A \to [[A \to (B \to C)] \to (B \to C)]$

t13. $[A \to [B \to (C \to D)]] \to [B \to [A \to (C \to D)]]$

t14. $A \to [(A \to B) \to B]$

t15. $[A \to (B \to C)] \to [B \to (A \to C)]$

t16. $(A \wedge B) \to [[A \to (B \to C)] \to C]$

t17. $(A \to B) \to [[A \wedge (B \to C)] \to C]$

t18. $[A \wedge (B \to C)] \to [(A \to B) \to C)]$

t19. $B \to [[A \to (B \to C)] \to (A \to C)]$

t20. $A \to (A \to A)$

t21. $A \to [B \to (A \vee B)]$

t22. $(A \to B) \vee (B \to A)$

t23. $[A \to (B \vee C)] \to [(A \to B) \vee (A \to C)]$

t24. $[(A \wedge B) \to C] \to [(A \to C) \vee (B \to C)]$

t25. $B \rightarrow (A \rightarrow A)$

t26. $(A \rightarrow B) \rightarrow [C \rightarrow (A \rightarrow B)]$

t27. $A \rightarrow (B \rightarrow A)$

t28. $A \rightarrow [B \rightarrow (C \rightarrow A)]$

t29. $(A \vee B) \rightarrow [(A \rightarrow B) \rightarrow B]$

t30. $A \rightarrow [B \rightarrow (A \wedge B)]$

t31. $[(A \wedge B) \rightarrow C] \rightarrow [A \rightarrow (B \rightarrow C)]$

t32. $(A \rightarrow B) \vee [(A \rightarrow B) \rightarrow C]$

t33. $A \vee (A \rightarrow B)$

t34. $A \vee [A \rightarrow [B \vee (B \rightarrow C)]]$

t35. $A \rightarrow [B \vee [(A \rightarrow B) \rightarrow C]]$

t36. $[A \vee (B \rightarrow C)] \vee (A \rightarrow B)$

t37. $(A \rightarrow C) \vee (B \rightarrow A)$

t38. $A \vee [B \rightarrow (A \rightarrow C)]$

t39. $B \rightarrow [(B \rightarrow C) \vee (A \rightarrow B)]$

t40. $B \rightarrow [A \vee (A \rightarrow C)]$

t41. $[(A \rightarrow B) \wedge \neg B] \rightarrow \neg A$

t42. $\neg B \rightarrow [(A \rightarrow B) \rightarrow \neg A]$

t43. $(A \rightarrow B) \rightarrow (\neg B \rightarrow \neg A)$

t44. $(A \rightarrow \neg B) \rightarrow (B \rightarrow \neg A)$

t45. $B \rightarrow [(A \rightarrow \neg B) \rightarrow \neg A]$

t46. $A \rightarrow [(A \rightarrow \neg B) \rightarrow \neg B]$

t47. $[A \rightarrow (B \rightarrow \neg C)] \rightarrow [B \rightarrow (A \rightarrow \neg C)]$

t48. $A \rightarrow \neg\neg A$

t49. $\neg(A \wedge \neg A)$

t50. $(A \rightarrow \neg A) \rightarrow \neg A$

t51. $A \rightarrow \neg(A \rightarrow \neg A)$

t52. $(A \rightarrow B) \rightarrow \neg(A \wedge \neg B)$

t53. $(A \wedge \neg B) \rightarrow \neg(A \rightarrow B)$

t54. $(A \rightarrow \neg B) \rightarrow \neg(A \wedge B)$

t55. $(A \wedge B) \rightarrow \neg(A \rightarrow \neg B)$

t56. $(A \rightarrow \neg B) \rightarrow [(A \rightarrow B) \rightarrow \neg A]$

t57. $(A \rightarrow B) \rightarrow [(A \rightarrow \neg B) \rightarrow \neg A]$

t58. $\neg(A \wedge B) \rightarrow (\neg A \vee \neg B)$

t59. $\neg A \vee \neg\neg A$

t60. $(A \rightarrow B) \vee \neg(A \rightarrow B)$

t61. $A \vee \neg A$

t62. $A \rightarrow [B \vee \neg(A \rightarrow B)]$

t63. $A \vee [A \rightarrow (B \vee \neg B)]$

t64. $(A \vee \neg B) \vee (A \rightarrow B)$

t65. $B \rightarrow (A \vee \neg A)$

t66. $\neg A \vee (B \rightarrow A)$

t67. $A \vee (B \rightarrow \neg A)$

t68. $\neg A \rightarrow (B \rightarrow \neg A)$

t69. $B \rightarrow [\neg B \vee (A \rightarrow B)]$

t70. $(A \wedge \neg A) \rightarrow B$

t71. $A \rightarrow (\neg A \rightarrow B)$

Postulates:

pt1. $Rabc \Rightarrow \exists x(Rabx \ \& \ Raxc)$

pt2. $R^2 abcd \Rightarrow \exists x(Rbcx \ \& \ Raxd)$

pt3. $R^2 abcd \Rightarrow \exists x(Racx \ \& \ Rbxd)$

pt4. $Raaa$

pt5. $Rabc \Rightarrow R^2 abbc$

pt6. $Rabc \Rightarrow R^2 baac$

pt7. $R^2 abcd \Rightarrow \exists x, y(Racx \ \& \ Rbcy \ \& \ Rxyd)$

pt8. $R^2 abcd \Rightarrow \exists x, y(Racx \ \& \ Rbcy \ \& \ Ryxd)$

pt9. $Rabc \Rightarrow R^2 abbc$

pt10. $\exists x \in Z \ Raxa$

  $[Za$ iff for all $b, c \in K, Rabc \Rightarrow \exists x \in O \ Rxbc]$

pt11. $\exists x \in O \ Raxa$

pt12. $R^2 abcd \Rightarrow R^2 bacd$

pt13. $R^3 abcde \Rightarrow R^3 acbde$

pt14. $Rabc \Rightarrow Rbac$

pt15. $R^2abcd \Rightarrow R^2acbd$

pt16. $Rabc \Rightarrow R^2baac$

pt17. $Rabc \Rightarrow \exists x(Rabx \ \& \ Rbxc)$

pt18. $Rabc \Rightarrow \exists x(Rbax \ \& \ Raxc)$

pt19. $R^2abcd \Rightarrow R^2bcad$

pt20. $Rabc \Rightarrow (a \leq c \ \text{or} \ b \leq c)$

pt21. $Rabc \Rightarrow (a \leq c \ \text{or} \ b \leq c)$

pt22. $(Rabc \ \& \ Rade \ \& \ a \in O) \Rightarrow (b \leq e \ \text{or} \ d \leq c)$

pt23. $(Rabc \ \& \ Rade) \Rightarrow \exists x[(Rabx \ \text{or} \ Radx) \ \& \ x \leq c \ \& \ x \leq e]$

pt24. $(Rabc \ \& \ Rade) \Rightarrow \exists x[(Raxc \ \text{or} \ Raxe) \ \& \ b \leq x \ \& \ d \leq x]$

pt25. $Rabc \Rightarrow b \leq c$

pt26. $R^2abcd \Rightarrow Racd$

pt27. $Rabc \Rightarrow a \leq c$

pt28. $R^2abcd \Rightarrow a \leq d$

pt29. $Rabc \Rightarrow (Rbac \ \& \ a \leq c)$

pt30. $Rabc \Rightarrow (a \leq c \ \& \ b \leq c)$

pt31. $R^2abcd \Rightarrow \exists x(Raxd \ \& \ b \leq x \ \& \ c \leq x)$

pt32. $(Rabc \ \& \ Rade \ \& \ a \in O) \Rightarrow Rdbc$

pt33. $(Rabc \ \& \ a \in O) \Rightarrow b \leq a$

pt34. $(Rabc \ \& \ Rcde \ \& \ a \in O) \Rightarrow (b \leq a \ \text{or} \ d \leq c)$

pt35. $Rabc \Rightarrow \exists x \in K(Rbax \ \& \ x \leq a)$

pt36. $(Rabc \ \& \ Rade \ \& \ a \in O) \Rightarrow (d \leq a \ \text{or} \ b \leq e)$

pt37. $(Rabc \ \& \ Rade \ \& \ a \in O) \Rightarrow b \leq e$

pt38. $(Rabc \ \& \ Rcde \ \& \ a \in O) \Rightarrow d \leq a$

pt39. $(Rabc \ \& \ Rade) \Rightarrow (a \leq e \ \text{or} \ b \leq e)$

pt40. $Rabc \Rightarrow b \leq a$

pt41. $(Rabc \ \& \ c \in S) \Rightarrow \exists x \in K \ \exists y \in S(Rabx \ \& \ Raxy)$

pt42. $(R^2abcd \ \& \ d \in S) \Rightarrow \exists x \in K \ \exists y \in S(Rbcx \ \& \ Raxy)$

pt43. $(R^2abcd \ \& \ d \in S) \Rightarrow \exists x \in K \ \exists y \in S(Racx \ \& \ Rbxy)$

pt44. $(R^2abcd \ \& \ d \in S) \Rightarrow \exists x \in S \ R^2acbx$

pt45. $(R^2abcd \ \& \ d \in S) \Rightarrow \exists x \in S \ R^2bcax$

pt46. $(R^2abcd \ \& \ d \in S) \Rightarrow \exists x \in S \ R^2bacx$

pt47. $(R^3 abcde \ \& \ e \in S) \Rightarrow \exists x \in S \ R^3 acbdx$

pt48. $(Rabc \ \& \ c \in S) \Rightarrow \exists x \in S \ Rbax$

pt49. $(Rabc \ \& \ a \in O \ \& \ c \in S) \Rightarrow \exists x \in S \ Rbbx$

pt50. $(Rabc \ \& \ c \in S) \Rightarrow \exists x \in S \ R^2 abbx$

pt51. $(Rabc \ \& \ c \in S) \Rightarrow \exists x \in S \ R^2 baax$

pt52. $(Rabc \ \& \ c \in S) \Rightarrow \exists x \in K \ \exists y \in S(Rabx \ \& \ Rbxy)$

pt53. $(Rabc \ \& \ c \in S) \Rightarrow \exists x \in K \ \exists y \in S(Rbax \ \& \ Raxy)$

pt54. $(Rabc \ \& \ c \in S) \Rightarrow \exists x \in S \ R^2 abbx$

pt55. $(Rabc \ \& \ c \in S) \Rightarrow \exists x \in S \ R^2 baax$

pt56. $(R^2 abcd \ \& \ d \in S) \Rightarrow \exists x, y \in K \ \exists z \in S(Racx \ \& \ Rbcy \ \& \ Rxyz)$

pt57. $(R^2 abcd \ \& \ d \in S) \Rightarrow \exists x, y \in K \ \exists z \in S(Racx \ \& \ Rbcy \ \& \ Ryxz)$

pt58. $(Rabc \ \& \ Rade \ \& \ c \in S \ \& \ e \in S) \Rightarrow \exists x \in K \ \exists y \in S(Raxy \ \&$
   $b \le x \ \& \ d \le x)$

pt59. $(Rabc \ \& \ Rade \ \& \ a \in O \ \& \ c \in S \ \& \ e \in S) \Rightarrow \exists x \in S \ Rdbx$

pt60. $(Rabc \ \& \ Rade \ \& \ a \in O \ \& \ e \in S) \Rightarrow \exists x \in S(Rdbx \ \& \ x \le c)$

pt61. $(Rabc \ \& \ a \in O \ \& \ c \in S) \Rightarrow b \le a$

pt62. $(Rabc \ \& \ c \in S) \Rightarrow \exists x \in K(Rbax \ \& \ x \le a)$

pt63. $(Rabc \ \& \ Rcde \ \& \ a \in O \ \& \ e \in S) \Rightarrow (b \le a \ \text{or} \ d \le c)$

pt64. $(Rabc \ \& \ Rade \ \& \ a \in O \ \& \ c \in S) \Rightarrow (d \le a \ \text{or} \ b \le e)$

pt65. $(Rabc \ \& \ c \in S) \Rightarrow b \le a$

pt66. $(Rabc \ \& \ Rade \ \& \ a \in O \ \& \ c \in S) \Rightarrow b \le e$

pt67. $(Rabc \ \& \ Rcde \ \& \ a \in O \ \& \ e \in S) \Rightarrow d \le a$

pt68. $(R^2 abcde \ \& \ d \in S) \Rightarrow \exists x \in S \ Racx$

pt69. $(Rabc \ \& \ Rade \ \& \ c \in S) \Rightarrow (a \le e \ \text{or} \ b \le e)$

pt70. $\exists x \in K \ Raax$

pt71. $\exists x \in K \ Rbax$

A.2 f-AXIOMS AND THEIR CORRESPONDING f-POSTULATES

Axioms:

$$t1_f. \ [(A \rightarrow B) \wedge (B \rightarrow f)] \rightarrow (A \rightarrow f)$$
$$= [(A \rightarrow B) \wedge \neg B] \rightarrow \neg A$$

$t2_f.\ (B \to f) \to [(A \to B) \to (A \to f)]$

$\quad = \neg B \to [(A \to B) \to \neg A]$

$t3_f.\ (A \to B) \to [(B \to f) \to (A \to f)]$

$\quad = (A \to B) \to (\neg B \to \neg A)$

$t4_f.\ [A \wedge (A \to f)] \to f$

$\quad = \neg(A \wedge \neg A)$

$t5_f.\ [A \to (A \to f)] \to (A \to f)$

$\quad = (A \to \neg A) \to \neg A$

$t6_f.\ A \to [[A \to (A \to f)] \to f]$

$\quad = A \to \neg(A \to \neg A)$

$t7_f.\ [A \to (B \to f)] \to [(A \to B) \to (A \to f)]$

$\quad = (A \to \neg B) \to [(A \to B) \to \neg A]$

$t8_f.\ (A \to B) \to [[A \to (B \to f)] \to (A \to f)]$

$\quad = (A \to B) \to [(A \to \neg B) \to \neg A]$

$t9_f.\ [A \to (B \to f)] \to [(A \wedge B) \to f)]$

$\quad = (A \to \neg B) \to \neg(A \wedge B)$

$t10_f.\ [(A \to A) \wedge (B \to B)] \to f] \to f$

$\quad = \neg\neg[(A \to A) \wedge (B \to B)]$

$t11_f.\ A \Rightarrow (A \to f) \to f$

$\quad = A \Rightarrow \neg\neg A$

$t12_f.\ A \to [[A \to (B \to f)] \to (B \to f)]$

$\quad = A \to [(A \to \neg B) \to \neg B]$

$t13_f.\ [A \to [B \to (C \to f)]] \to [B \to [A \to (C \to f)]]$

$\quad = [A \to (B \to \neg C) \to [B \to (A \to \neg C)]$

$t14_f.\ A \to [(A \to f) \to f]$

$\quad = A \to \neg\neg A$

$t15_f.\ [A \to (B \to f)] \to [B \to (A \to f)]$

$\quad = (A \to \neg B) \to (B \to \neg A)$

$t16_f.\ (A \wedge B) \to [[A \to (B \to f)] \to f]$

$\quad = (A \wedge B) \to \neg(A \to \neg B)$

$t17_f.\ (A \to B) \to [[A \wedge (B \to f)] \to f]$

$\quad = (A \to B) \to \neg(A \wedge \neg B)$

$t18_f. \ [A \wedge (B \rightarrow f)] \rightarrow [(A \rightarrow B) \rightarrow f]$

$\quad = (A \wedge \neg B) \rightarrow \neg(A \rightarrow B)$

$t19_f. \ B \rightarrow [[A \rightarrow (B \rightarrow f)] \rightarrow (A \rightarrow f)]$

$\quad = B \rightarrow [(A \rightarrow \neg B) \rightarrow \neg A]$

$t20_f. \ f \rightarrow (f \rightarrow f)$

$\quad = f \rightarrow \neg f$

$t22_f. \ (A \rightarrow f) \vee (f \rightarrow A)$

$\quad = \neg A \vee (f \rightarrow A)$

$t24_f. \ [(A \wedge B) \rightarrow f] \rightarrow [(A \rightarrow f) \vee (B \rightarrow f)]$

$\quad = \neg(A \wedge B) \rightarrow (\neg A \vee \neg B)$

$t25_f. \ A \rightarrow (f \rightarrow f)$

$\quad = A \rightarrow \neg f$

$t26_f. \ (A \rightarrow f) \rightarrow [B \rightarrow (A \rightarrow f)]$

$\quad = \neg A \rightarrow (B \rightarrow \neg A)$

$t27_f. \ f \rightarrow (A \rightarrow f)$

$\quad = f \rightarrow \neg A$

$t28_f. \ f \rightarrow [B \rightarrow (C \rightarrow f)]$

$\quad = f \rightarrow (B \rightarrow \neg C)$

$t29_f. \ (A \vee f) \rightarrow [(A \rightarrow f) \rightarrow f]$

$\quad = (A \vee f) \rightarrow \neg\neg A$

$t32_f. \ (A \rightarrow B) \vee [(A \rightarrow B) \rightarrow f]$

$\quad = (A \rightarrow B) \vee \neg(A \rightarrow B)$

$t33_f. \ A \vee (A \rightarrow f)$

$\quad = A \vee \neg A$

$t34_f. \ A \vee [A \rightarrow [B \vee (B \rightarrow f)]]$

$\quad = A \vee [A \rightarrow (B \vee \neg B)]$

$t35_f. \ A \rightarrow [B \vee [(A \rightarrow B) \rightarrow f]]$

$\quad = A \rightarrow [B \vee \neg(A \rightarrow B)]$

$t36_f. \ [A \vee (B \rightarrow f)] \vee (A \rightarrow B)$

$\quad = (A \vee \neg B) \vee (A \rightarrow B)$

$t37_f. \ (A \rightarrow f) \vee (B \rightarrow A)$

$\quad = \neg A \vee (B \rightarrow A)$

$t38_f.\ A \vee [B \to (A \to f)]$

$\quad = A \vee (B \to \neg A)$

$t39_f.\ B \to [(B \to f) \vee (A \to B)]$

$\quad = B \to [\neg B \to \vee (A \to B)]$

Postulates:

$pt1_f.$ *Rabc & $c \in S \Rightarrow \exists x \in K\ \exists y \in S(Rabx\ \&\ Raxy)$*

$pt2_f.$ $R^2abcd \Rightarrow \exists x \in K\ \exists y \in S(Rbcx\ \&\ Raxy)$

$pt3_f.$ $R^2abcd \Rightarrow \exists x \in K\ \exists y \in S(Racx\ \&\ Rbxy)$

$pt4_f.$ *$a \in S \Rightarrow \exists x \in S\ Raax$*

$pt5_f.$ *(Rabc & $c \in S) \Rightarrow \exists x \in S\ R^2abbx$*

$pt6_f.$ *(Rabc & $c \in S) \Rightarrow \exists x \in S\ R^2baax$*

$pt7_f.$ *(R^2abcd & $d \in S) \Rightarrow \exists x, y \in K\ \exists z \in S(Racx\ \&\ Rbcy\ \&\ Rxyz)$*

$pt8_f.$ *(R^2abcd & $d \in S) \Rightarrow \exists x, y \in K\ \exists z \in S(Racx\ \&\ Rbcy\ \&\ Ryxz)$*

$pt9_f.$ *(Rabc & $c \in S) \Rightarrow \exists x \in S\ R^2abbx$*

$pt10_f.$ *$a \in S \Rightarrow \exists x \in Z\ \exists y \in S\ Raxy$*

$\quad$ [*Za* iff for all $b, c \in K$, *Rabc* $\Rightarrow \exists x \in O\ Rxbc$]

$pt11_f.$ *$a \in S \Rightarrow \exists x \in O\ \exists y \in S\ Raxy$*

$pt12_f.$ *(R^2abcd & $d \in S) \Rightarrow \exists x \in S\ R^2bacx$*

$pt13_f.$ *(R^3abcde & $e \in S) \Rightarrow \exists x \in S\ R^3acbdx$*

$pt14_f.$ *(Rabc & $c \in S) \Rightarrow \exists x \in S\ Rbax$*

$pt15_f.$ *(R^2abcd & $d \in S) \Rightarrow \exists x \in S\ R^2acbx$*

$pt16_f.$ *(Rabc & $c \in S) \Rightarrow \exists x \in S\ R^2baax$*

$pt17_f.$ *(Rabc & $c \in S) \Rightarrow \exists x \in K\ \exists y \in S(Rabx\ \&\ Rbxy)$*

$pt18_f.$ *(Rabc & $c \in S) \Rightarrow \exists x \in K\ \exists y \in S(Rbax\ \&\ Raxy)$*

$pt19_f.$ *(R^2abcd & $d \in S) \Rightarrow \exists x \in S\ R^2bacx$*

$pt20_f.$ *(Rabc & $c \in S) \Rightarrow (a \in S\ or\ b \in S)$*

$pt22_f.$ *(Rabc & Rade & $a \in O$ & $c \in S) \Rightarrow (d \in S\ or\ b \leq e)$*

$pt24_f.$ *(Rabc & Rade & $c \in S$ & $d \in S) \Rightarrow \exists x \in K\ \exists y \in S(Raxy\ \&$*

$\quad b \leq x\ \&\ d \leq x)$

pt25$_f$. $(Rabc \ \& \ c \in S) \Rightarrow b \in S$

pt26$_f$. $(R^2 abcd \ \& \ d \in S) \Rightarrow \exists x \in S \ Racx$

pt27$_f$. $(Rabc \ \& \ c \in S) \Rightarrow a \in S$

pt28$_f$. $(R^2 abcd \ \& \ d \in S) \Rightarrow a \in S$

pt29$_f$. $(Rabc \ \& \ c \in S) \Rightarrow (a \in S \ \& \ \exists x \in S \ Rbax)$

pt32$_f$. $(Rabc \ \& \ Rade \ \& \ a \in O \ \& \ e \in S) \Rightarrow Rdbc$

pt33$_f$. $(Rabc \ \& \ a \in O \ \& \ c \in S) \Rightarrow b \leq a$

pt34$_f$. $(Rabc \ \& \ Rcde \ \& \ a \in O \ \& \ e \in S) \Rightarrow (b \leq a \ \text{or} \ d \leq c)$

pt35$_f$. $(Rabc \ \& \ c \in S) \Rightarrow \exists x \in K(Rbax \ \& \ x \leq a)$

pt36$_f$. $(Rabc \ \& \ Rade \ \& \ a \in O \ \& \ c \in S) \Rightarrow (d \leq a \ \text{or} \ b \leq e)$

pt37$_f$. $(Rabc \ \& \ Rade \ \& \ a \in O \ \& \ c \in S) \Rightarrow b \leq e$

pt38$_f$. $(Rabc \ \& \ Rcde \ \& \ a \in O \ \& \ e \in S) \Rightarrow d \leq a$

pt39$_f$. $(Rabc \ \& \ Rade \ \& \ c \in S) \Rightarrow (a \leq e \ \text{or} \ b \leq e)$

BIBLIOGRAPHY

[1] W. Ackermann, Begründung einer strengen implikation, Journal of Symbolic Logic 21 (2) (1956) 113–128.

[2] A.R. Anderson, N.D. Belnap Jr., Entailment. The Logic of Relevance and Necessity, vol I, Princeton University Press, Princeton, NJ, 1975.

[3] A.R. Anderson, N.D. Belnap Jr., J.M. Dunn, Entailment: The Logic of Relevance and Necessity, vol. II, Princeton University Press, Princeton, NJ, 1992.

[4] H. Andréka, J.X. Madarász, I. Németi, Mutual definability does not imply definitional equivalence, a simple example, Mathematical Logic Quarterly 51 (6) (2005) 591–597.

[5] J. Beall, R.T. Brady, J.M. Dunn, A.P. Hazen, E. Mares, R.K. Meyer, G. Priest, G. Restall, D. Ripley, J. Slaney, R. Sylvan, On the ternary relation and conditionality, Journal of Philosophical Logic 41 (3) (2012) 595–612.

[6] K. Bimbó, The third place is the charm: the emergence, the development and the future of the ternary relational semantics for relevance and some other non-classical logics, Research project funded by an Insight Grant awarded by the Social Sciences and Humanities Research Council of Canada, https://www.ualberta.ca/research/research-initiatives/the-third-place-is-the-charm, 2015–2018.

[7] K. Bimbó, J.M. Dunn, Generalized Galois Logics. Relational Semantics of Nonclassical Logical Calculi, CSLI Lecture Notes, vol. 188, CSLI, Stanford, CA, 2008.

[8] K. Bimbó, J.M. Dunn (Eds.), Proceedings of the Third Workshop, 16–17 May 2016, Edmonton, Canada, IfCoLog Journal of Logics and their Applications 4 (3) (2017).

[9] R.T. Brady (Ed.), Relevant Logics and Their Rivals, vol. II, Ashgate, Aldershot, 2003.

[10] R.T. Brady, Universal Logic, CSLI, Stanford, CA, 2006.

[11] W. Carnielli, M.E. Coniglio, J. Marcos, Logics of formal inconsistency, in: Handbook of Philosophical Logic, vol. 14, Springer, Dordrecht, Netherlands, 2007, pp. 1–93.

[12] B.J. Copeland, On when a semantics is not a semantics: some reasons for disliking the Routley-Meyer semantics for relevance logic, Journal of Philosophical Logic 8 (1) (1979) 399–413.

[13] K. Došen, Negation in the light of modal logic, in: What Is Negation?, in: Applied Logic Series, vol. 13, Springer, Netherlands, 1999, pp. 77–86.

[14] K. Došen, Negation as a modal operator, Reports on Mathematical Logic 20 (1986) 15–27.

[15] K. Fine, Models for entailment, Journal of Philosophical Logic 3 (4) (1974) 347–372.

[16] D.M. Gabbay, F. Guenthner (Eds.), Handbook of Philosophical Logic, 2nd ed., Kluwer Academic Publishers, 2002.

[17] C. González, MaTest, http://ceguel.es/matest, 2012. (Accessed 30 May 2017).

[18] I. Hacking, What is strict implication?, Journal of Symbolic Logic 28 (1) (1963) 51–71.

[19] J.M. Méndez, Constructive R, Bulletin of the Section of Logic 16 (1987) 167–175.

[20] J.M. Méndez, G. Robles, Relevance logics and intuitionistic negation, Journal of Applied Non-Classical Logics 18 (1) (2008) 49–65.

[21] S.P. Odintsov, Constructive Negations and Paraconsistency, Trends in Logic Series, vol. 26, Springer, Dordrecht, Netherlands, 2008.

[22] G. Restall, Subintuitionistic logics, Notre Dame Journal of Formal Logic 35 (1) (1994) 116–129.

[23] G. Restall, Negation in relevant logics (How I stopped worrying and learned to love the Routley Star), in: What Is Negation?, in: Applied Logic Series, vol. 13, Springer, Netherlands, 1999, pp. 53–76.

[24] G. Robles, Negaciones subintuicionistas para lógicas con la Conversa de la Propiedad Ackermann, Ediciones Universidad de Salamanca, ISBN 84-7800-468-8, 2006.

[25] G. Robles, Relevance logics and intuitionistic negation II. Negation introduced with the unary connective, Journal of Applied Non-Classical Logics 19 (3) (2009) 371–388.

[26] G. Robles, Minimal non-relevant logics without the K axiom II. Negation introduced with the unary connective, Reports on Mathematical Logic 45 (2010) 97–118.

[27] G. Robles, J.M. Méndez, Minimal non-relevant logics without the K axiom, Reports on Mathematical Logic 42 (2007) 117–144.

[28] G. Robles, J.M. Méndez, Strong paraconsistency and the basic constructive logic for an even weaker sense of consistency, Journal of Logic, Language and Information 18 (3) (2009) 357–402.

[29] G. Robles, J.M. Méndez, A binary Routley semantics for intuitionistic De Morgan minimal logic H_M and its extensions, Logic Journal of the IGPL 23 (2) (2015) 174–193.

[30] G. Robles, J.M. Méndez, The minimal constructive logic determined by binary relational semantics, in preparation.

[31] G. Robles, J.M. Méndez, F. Salto, Minimal negation in the ternary relational semantics, Reports on Mathematical Logic 39 (2005) 47–65.

[32] G. Robles, J.M. Méndez, F. Salto, Relevance logics, paradoxes of consistency and the K rule, Logique et Analyse 198 (2007) 129–145.

[33] R. Routley, R.K. Meyer, The semantics of entailment I, in: H. Leblanc (Ed.), Truth, Syntax and Modality. Proceedings of the Temple University Conference on Alternative Semantics, in: Studies in Logic and the Foundations of Mathematics, vol. 68, North-Holland Publishing Company, Amsterdam and London, 1973, pp. 199–243.

[34] R. Routley, R.K. Meyer, The semantics of entailment II, Journal of Philosophical Logic 1 (1) (1972) 53–73.

[35] R. Routley, R.K. Meyer, The semantics of entailment III, Journal of Philosophical Logic 1 (2) (1972) 192–208.

[36] R. Routley, V. Routley, The semantics of first degree entailment, Noûs 6 (4) (1972) 335–359.

[37] R. Routley, R.K. Meyer, V. Plumwood, R.T. Brady, Relevant Logics and Their Rivals, vol. 1, Ridgeview Publishing Co., Atascadero, CA, 1982.

[38] Y.V. Shramko, Relevant variants of intuitionistic logic, Logic Journal of IGPL 2 (1) (1994) 47–53.

[39] J. Slaney, MaGIC, Matrix Generator for Implication Connectives: Version 2.1, Notes and Guide, Australian National University, 1995, http://users.cecs.anu.edu.au/~jks/magic.html, 1995. (Accessed 31 May 2017).

[40] R. Sylvan, V. Plumwood, Non-normal relevant logics, in: R. Brady (Ed.), Relevant Logics and Their Rivals, vol. II, in: Western Philosophy Series, Ashgate, Aldershot and Burlington, 2003, pp. 10–16.

[41] N. Tennant, Anti-Realism and Logic: Truth as Eternal, Clarendon Press, 1987.

[42] A. Urquhart, Semantics for relevant logics, Journal of Symbolic Logic 37 (1) (1972) 159–169.

[43] E. Yang, R and relevance principle revisited, Journal of Philosophical Logic 42 (5) (2013) 767–782.
[44] E.N. Zalta (Ed.), The Stanford Encyclopedia of Philosophy, The Metaphysics Research Lab Center for the Study of Language and Information, Stanford University, Stanford, CA, 2016, https://plato.stanford.edu/.

INDEX